Basic Laws of Th
Equilibrium Radiation and Their Role in Thermodynamics

Chen Dayou

AMERICAN ACADEMIC PRESS

AMERICAN ACADEMIC PRESS

By AMERICAN ACADEMIC PRESS

201 Main Street

Salt Lake City

UT 84111 USA

Email manu@AcademicPress.us

Visit us at http://www.AcademicPress.us

ISBN: 979-8-3370-8958-4

Distributed to the trade by National Book Network Suite 200, 4501 Forbes Boulevard, Lanham, MD 20706

10 9 8 7 6 5 4 3 2 1

Abstract

Employing classical statistical theory as a tool, this paper conducts an in-depth study on the fundamental laws of thermal equilibrium radiation and their decisive role in the second law of thermodynamics. The basic laws of thermal equilibrium radiation encompass both the energy distribution of oscillators and the energy exchange between oscillators and the radiation field. Through research on blackbody radiation laws, Planck's formula, and the characteristics of radiative absorption by oscillators, this paper reveals the principles governing oscillator excitation and radiation absorption. Regarding the fundamental laws of thermal equilibrium radiation, we provide a comprehensive formulation: material oscillators in a thermal equilibrium radiation field obey the Boltzmann energy distribution law; the probability of radiative excitation of an oscillator is proportional to its energy level state; the probability of radiative absorption by an oscillator is proportional to its energy distribution probability. The former reveals the law of energy distribution from a static perspective, while the latter reveals the law of energy exchange between oscillators and the radiation field from a dynamic perspective. Together, they constitute the basic laws of thermal radiation. Based on these fundamental laws of thermal radiation, the microscopic dynamics and characteristics of thermal equilibrium processes are analyzed. The relationship between the basic laws of thermal radiation and the second law of thermodynamics is examined, and the necessary and sufficient conditions for thermal equilibrium radiation are proposed and demonstrated. It is revealed that the fundamental laws of thermal radiation are the decisive factor for the second law of thermodynamics, leading to the conclusion that the second law of thermodynamics is the macroscopic manifestation of these fundamental laws.

In the appendix of this paper, we propose and demonstrate an analytical expression for Planck's constant, explore and analyze the elastic mechanical properties of quantum space, and, on this basis, investigate foundational issues of quantum mechanics. Starting from the discontinuous nature of microscopic particle motion, the de Broglie relation is derived, and the microscopic dynamics of radiative excitation by harmonic oscillators are deduced, revealing the dynamical characteristics of quantum mechanics. In the context of macroscopic physics, the concept of volume energy is introduced, and foundational issues of general relativity are explored. It is proposed and argued that the elastic potential energy density of quantum space corresponds to the components of the energy-momentum tensor of the gravitational field. By directly solving Einstein's field equations using the variational principle of fields, the complete solution of the Schwarzschild metric is obtained. The quantum spatial significance of the mass-energy relationship is explored, unveiling the dynamical characteristics of cosmic expansion. The appendix investigates foundational and cutting-edge issues in physics, serving as an extension of the main text.

Introduction

The theoretical research and groundbreaking experiments on thermal radiation hold significant importance in the history of physics. By the early 20th century, a series of theoretical and experimental achievements had been made in the field of thermal radiation, including Kirchhoff's law, Wien's displacement law, the Stefan-Boltzmann law, and Planck's blackbody radiation law. In particular, during the revelation of the blackbody radiation law, Planck abandoned the conventional notion of energy continuity and introduced the concept of energy quanta. This not only successfully resolved the so-called "ultraviolet catastrophe" that had perplexed the theoretical community but also marked the dawn of quantum physics.

So why do we raise the proposition of the fundamental laws of thermal radiation? While thermodynamics has its three fundamental laws, thermal radiation also has Wien’s law and Planck’s formula. But are there even more fundamental laws governing thermal radiation? The answer is yes. The basic laws of thermal radiation are implicitly embedded within the blackbody radiation law and the

Rayleigh-Jeans formula. By carefully unraveling these complexities and uncovering their mysteries, we can gain new insights and reveal the fundamental laws of thermal equilibrium radiation.

The fundamental laws of thermal equilibrium radiation encompass two aspects: the energy distribution law of microscopic material components and the law governing energy exchange between material oscillators and the radiation field. The former is given by the Boltzmann energy distribution law, while the latter is the focus of our research. The energy exchange between oscillators and the radiation field can be further divided into two levels: excitation and absorption of radiation. After in-depth study, our conclusion regarding the law of energy exchange between material oscillators and the radiation field is as follows:

· The probability of an oscillator exciting radiation is proportional to its energy level state.

· The probability of an oscillator absorbing radiation is proportional to its energy distribution probability.With the specific form of the energy exchange law between microscopic oscillators and the radiation field, the complete description of the fundamental law of thermal equilibrium radiation is as follows: in a thermal equilibrium radiation field, material oscillators obey the Boltzmann energy distribution law in terms of energy distribution; the probability of an oscillator exciting radiation is proportional to its energy level state; and the probability of an oscillator absorbing radiation is proportional to its energy distribution probability. The former reveals the energy distribution law of oscillators from a static perspective, while the latter dynamically reveals the law of energy exchange between material oscillators and the radiation field. Together, they are referred to as the fundamental laws of thermal equilibrium radiation.

Interestingly, the Boltzmann energy distribution law can be directly derived from the energy exchange law between oscillators and the radiation field. This is why we include the Boltzmann energy distribution law as part of the fundamental laws of thermal radiation. Through the analysis of thermal equilibrium processes, we propose the idea that in the dynamic equilibrium process of an independent thermodynamic system, there exists a fixed point of thermodynamic temperature. We also assert that the second law of thermodynamics is the macroscopic manifestation of the fundamental laws of thermal radiation.

In the appendix of this book, the derivation and proof of the analytical expression of Planck's constant involve the physical properties of space. The appendix introduces the concept of a quantized elastic space and, on this basis, explores and analyzes foundational and cutting-edge issues in quantum mechanics and general relativity. It can serve as extended reading material to complement the main text.

Author

Contents

Abstract I

Introduction IV

S.1 Basic Concepts and Theoretical Achievements of Thermal Radiation Theory 1

S.2 Planck's Black-body Radiation Law 23

S.3 Laws of Energy Exchange Between Oscillators and the Radiation Field 40

S.4 Fundamental Laws of Thermal Radiation 54

S.5 Analysis of Thermal Equilibrium Radiation Process 64

S.6 The Fundamental Laws of Thermal Radiation and the Second Law of Thermodynamics 75

S.7 Description of the Thermal Equilibrium Radiation Process State 88

S.8 Conclusion 96

Appendix: Derivation and Proof of the Analytical Expression for Planck's Constant 99

S.1 The Elastic Mechanical Properties of Quantum Space.... 99

S.2 Volume Energy and Particle Transient Gravitational Field Fluctuations 120

S.3 Mechanism of Mass Generation; Relationship Between Fundamental Length of Space and Nuclear Density Constant;

Quantum Space Background of Newtonian Mechanics 130
S.4 Quantum Space's Limit Amplitude; Derivation of de Broglie Relation; Analysis of Harmonic Oscillator Excited Radiation; Significance of Planck's Constant 145
S.5 Mass-Energy Relation and Dark Energy; Gravitational Field Energy-Momentum Tensor; Einstein Field Equation Variational Solution; Dynamics of Cosmic Expansion 175
References 197

S.1 Basic Concepts and Theoretical Achievements of Thermal Radiation Theory

1.1 Kirchhoff's Law and Black Body Radiation

Modern radiation theory asserts that thermal radiation consists of electromagnetic waves with specific wavelengths, excited by changes in the internal energy of an object. The internal energy of an object is represented by its macroscopic temperature. As long as the thermodynamic temperature of the object is not absolute zero, it radiates electromagnetic waves, emitting thermal energy into the surrounding space. At the same time, it also absorbs electromagnetic waves from the surrounding space, thereby absorbing thermal energy. Due to the complex motion of material oscillators, different wavelengths of electromagnetic waves are excited. In summary, it can be considered that the thermal energy radiated by an object is a function of temperature T and wavelength λ (or frequency ν). To quantitatively study the intensity of the thermal energy radiated by an object, we typically examine the spectral energy density of radiation per unit surface area of the object.

1) Radiative Intensity, Irradiance, and Absorption Coefficient

When the temperature of an object is T, the energy radiated by the object from its unit surface area per unit time across various wavelengths λ is called the radiative emissivity of the object and is denoted as $R(\lambda,T)$. The energy radiated per unit time from the unit surface area of the object across wavelengths from λ to $\lambda + d\lambda$ is $R(\lambda,T)$. When $d\lambda$ is sufficiently small, it is directly proportional to $d\lambda$. The ratio is defined as

$$r_{\lambda}(\lambda,T) = \frac{dR(\lambda,T)}{d\lambda} \tag{1-1}$$

as the monochromatic emissivity of the object at wavelength λ, also called the monochromatic radiative ability. Clearly, we also have

$$R(T) = R(\lambda,T) = \int_0^{\infty} r_{\lambda}(\lambda,T) d\lambda \tag{1-2}$$

We can also use the same method to describe the radiation energy irradiated onto the unit surface area of an object by the radiation field. When the temperature is T, the energy irradiated onto the unit surface area of the object per unit time across various wavelengths λ is called the spectral irradiance, denoted as $E(\lambda,T)$. The energy λ within the wavelength range from $\lambda + d\lambda$

to $\lambda + d\lambda$ is proportional to $dE(\lambda, T)$, and the ratio is called the monochromatic spectral irradiance, denoted as

$$e_\lambda(\lambda, T) = \frac{dE(\lambda, T)}{d\lambda} \tag{1-3}$$

At the same time, we also have

$$E(T) = E(\lambda, T) = \int_0^\infty e_\lambda(\lambda, T) d\lambda \tag{1-4}$$

Empirical and experimental evidence shows that, in general, an object cannot absorb all the radiation that hits its surface. However, it can be assumed that when $d\lambda$ is sufficiently small, the energy λ absorbed by the object across wavelengths from $\lambda + d\lambda$ to $dE'(\lambda, T)$ is not only proportional to $d\lambda$, but also proportional to the irradiance $e_\lambda(\lambda, T)$. That is, $dE'(\lambda, T)$ is proportional to the product $e_\lambda(\lambda, T) d\lambda$. The ratio is defined as

$$\alpha(\lambda, T) = \frac{dE'(\lambda, T)}{dE(\lambda, T)} = \frac{dE'(\lambda, T)}{e_\lambda(\lambda, T) d\lambda} \tag{1-5}$$

T as the monochromatic absorption coefficient of the object for radiation energy in the wavelength range from λ to $\lambda + d\lambda$, also called the spectral absorption coefficient.

2) Kirchhoff's Law

The monochromatic radiation emissivity $r_\lambda(\lambda, T)$ and monochromatic absorption coefficient $\alpha(\lambda, T)$ introduced above are characteristic quantities that describe the radiative properties of an object. Different objects have different values for these quantities. Although these quantities vary with the type of object, they have an intrinsic relationship. A wealth of experimental evidence indicates that the stronger an object's ability to absorb radiation, the greater its ability to emit radiation. Conversely, if an object has a weaker ability to absorb radiation, its ability to emit radiation will be weaker as well. For instance, under sunlight, a black object heats up faster than a white object, and after the sunlight is blocked, the black object cools down faster than the white object. This is because the absorption coefficient $\alpha(\lambda, T)$ of the black object is higher compared to that of the white object. Furthermore, a steel block heated to 800°C emits a bright red light, while molten quartz at the same temperature does not emit light because quartz is a transparent crystal and has a much lower ability to absorb light than steel, resulting in a lower emission ability. A large amount of experimental data reveals that the monochromatic radiation emissivity $r_\lambda(\lambda, T)$ of an object is directly proportional to its monochromatic absorption coefficient $\alpha(\lambda, T)$ The in-depth research of German physicist G.H. Kirchhoff further revealed that the ratio of these two characteristic quantities for different objects is actually equal. That is, under the same temperature conditions, we have

$$\frac{r_{\lambda 1}}{\alpha_1} = \frac{r_{\lambda 2}}{\alpha_2} = \ldots = \frac{r_{\lambda n}}{\alpha_n} \tag{1-6}$$

This important and intriguing relationship can be demonstrated with the help of the thermal equilibrium process of an independent cavity system. Assume there is a large vacuum container, inside which there are several objects made of different materials and at different temperatures. Assume that the cavity is isolated from external heat sources. In this case, the container and the objects inside it can only exchange energy through emission and absorption of radiation. We know that after a sufficiently long period of time, the container and the radiation field within it, along with the objects, will tend to reach a common temperature. At this point, the radiation field within the container will exhibit isotropic characteristics. The energy density and the energy spectral density $e_\lambda(\lambda,T)$ will be the same throughout the radiation field. It can be proven that the relationship between the monochromatic irradiance $e_\lambda(\lambda,T)$ in the cavity and the energy spectral density $u_\lambda(\lambda,T)$ is:

$$e_\lambda(\lambda,T) = \frac{c}{4} u_\lambda(\lambda,T) \tag{1-7}$$

From this, we can deduce that the monochromatic irradiance $e_\lambda(\lambda,T)$ in the radiation field of the container is the same everywhere. Under these conditions, in a state of thermal equilibrium, the radiation energy absorbed by each object at different wavelengths should be equal to the radiation energy it emits at the corresponding wavelengths. Therefore, we have the following equation:

$$r_{i\lambda}(\lambda,T)d\lambda = \alpha_i(\lambda,T)e_{i\lambda}(\lambda,T) \ (i=1,2,\ldots,n) \qquad (1\text{-}8)$$

or

$$\frac{r_{i\lambda}(\lambda,T)}{\alpha_i(\lambda,T)} = e_{i\lambda}(\lambda,T) \ (i=1,2,\ldots,n) \qquad (1\text{-}9)$$

Under thermal equilibrium conditions, the monochromatic irradiance at all points in the radiation field is the same. At this point, equation (1-6) can be further written as:

$$\frac{r_{1\lambda}}{\alpha_1} = \frac{r_{2\lambda}}{\alpha_2} = \ldots = \frac{r_{n\lambda}}{\alpha_n} = e_\lambda(\lambda,T) \qquad (1\text{-}10)$$

Equation (1-10) is the famous Kirchhoff's Law of thermal radiation.

We can discuss Kirchhoff's Law in more detail. At first glance, it may not be surprising if the ratio of the monochromatic emissivity to the monochromatic absorption coefficient of objects made from different materials differs, as materials vary. However, contrary to intuition, the ratio of these two characteristic quantities for different materials is exactly the same. This law can be explained using thermodynamic principles. We continue with the example of cavity thermal equilibrium radiation, where $e_\lambda(\lambda,T)$ represents the monochromatic irradiance of the radiation field in the cavity impinging on the unit surface area of an object, exhibiting isotropic characteristics. Suppose that the monochromatic irradiance $e_{i\lambda}(\lambda,T)$ at two locations in the cavity is different. If the irradiance at location A is greater than that at location B ($e_{1\lambda}(\lambda,T) \triangleright e_{2\lambda}(\lambda,T)$, we can place two heat-absorbing bodies at locations A and B. Because of the differing irradiance, the temperature at location A will gradually increase compared to location B, creating two heat sources at different temperatures. This temperature difference can then be used to build a heat engine that continuously produces work without any change in the system. However, we know that such a perpetual motion machine would violate the second law of thermodynamics, which negates the possibility of anisotropic irradiance in a thermal equilibrium radiation field. This is where Kirchhoff's insight lies.

In general, the absorption coefficient $\alpha(\lambda,T)$ of an object's thermal radiation for any wavelength of radiation is always less than 1. This situation indicates that no object's emissivity $R(\lambda,T)$ is equal to the instantaneous irradiance $E(\lambda,T)$ projected onto it.

This is because every object has a certain reflection capability. We know that black objects have a large absorption coefficient $\alpha(\lambda,T)$, and objects coated with black soot or enamel can absorb almost 99% of sunlight. Thus, the ability of an object to absorb radiation depends only on the surface properties of the object, and not on other physical properties. Hence, it was postulated whether it is possible to find an object whose $\alpha(\lambda,T)=1$, at any temperature T, can completely absorb radiation of any wavelength λ projected onto it. Such an ideal object is called a perfect black body. Of course, this is only an idealized model and does not actually exist. However, one can imagine a closed cavity container with a small hole S in the wall, as shown in figure (1-1).

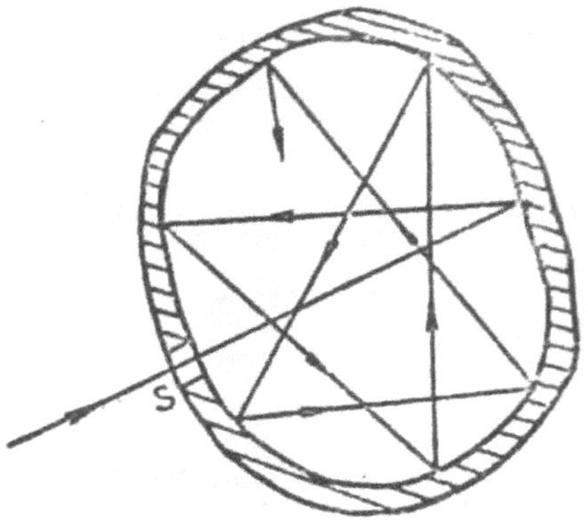

Figure (1-1) Absolute Black Body—Cavity

In this case, the radiation entering the container through the small hole S from the outside, after multiple reflections on the inner wall of the container, is almost entirely absorbed by the container, with very little possibility of escaping through the hole. Thus, the absorption coefficient of the small hole S in the cavity can be approximately considered as 1, and the small hole S in the cavity effectively behaves as the surface of a perfect black body. The radiative characteristics of the small hole depend only on the thermodynamic temperature of the cavity and are independent of the material of the cavity.

Indeed, the small hole S also emits radiation, and this emission is independent of the container's material. If the temperature of the container is T, the monochromatic emissivity of the absolute black-body is $r_{B\lambda}(\lambda,T)$, then equation (1-10) can be extended to:

$$\frac{r_{1\lambda}}{\alpha_1}=\frac{r_{2\lambda}}{\alpha_2}=\ldots=\frac{r_{n\lambda}}{\alpha_n}=\frac{r_{B\lambda}}{1}=e_\lambda(\lambda,T)=r_{B\lambda}(\lambda,T) \qquad (1\text{-}11)$$

Equation (1-11) is also referred to as Kirchhoff's Law. Since $r_{B\lambda}(\lambda,T)$ is measurable, once the emissivity of the black body $r_{B\lambda}(\lambda,T)$ is known, it is not difficult to determine the radiative properties of any object. For example, $r_{n\lambda}(\lambda,T)=\alpha_n(\lambda,T)r_{B\lambda}(\lambda,T)$, by knowing either the emissivity or the absorption coefficient, we can fully understand the radiative characteristics of the object n. Thus, the study of cavity radiation becomes a popular topic in the field of thermal equilibrium radiation.

For a cavity container with a small hole, where the opening S behaves like the surface of an absolute black body, we can obtain the monochromatic emissivity of the absolute black body at temperature T by controlling the thermodynamic temperature of the cavity to T. The cavity radiation experiment is shown in Figure (1-2).

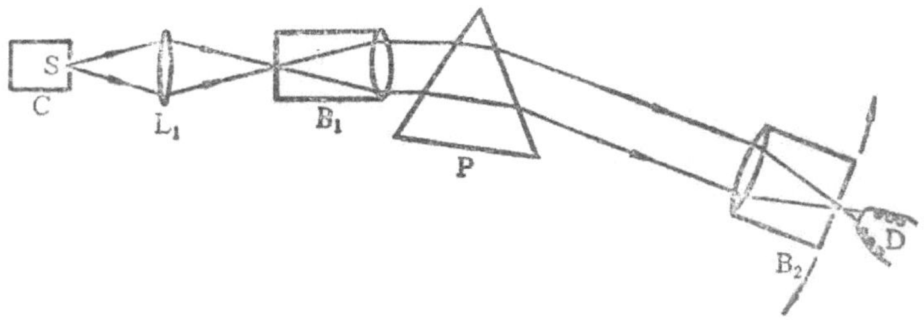

Figure (1-2) Schematic Diagram of the Cavity Radiation Experimental Setup

In the diagram, C represents a cavity container with a specific temperature, and S is a narrow aperture with a known area cut into the container wall. The radiation emitted from the narrow aperture passes through the lens L and is converged onto the parallel light tube B_1, forming parallel light that is incident on the prism P. Utilizing the dispersive function of the prism, radiation of different wavelengths is emitted in different directions after passing through the prism, achieving the function of spectral separation. By aligning the parallel light tube B_2 with different orientations of the prism, radiation of different wavelengths can be focused onto the radiation energy detector D, thereby measuring the monochromatic emissivity λ of the cavity radiation at different wavelengths $r_{B\lambda}(\lambda,T)$. After the experimental data is completed, the absolute black-body thermal radiation energy spectrum curve for a temperature of T can be plotted. By adjusting the temperature of the cavity container, the absolute black-body radiation energy spectrum curves for different temperature conditions can be obtained. At the end of the 19th century, the renowned experimental physicists O.R. Lummer and E. Pringsheim collaborated to accurately measure the absolute black-body energy spectrum curves at different temperatures for the first time.

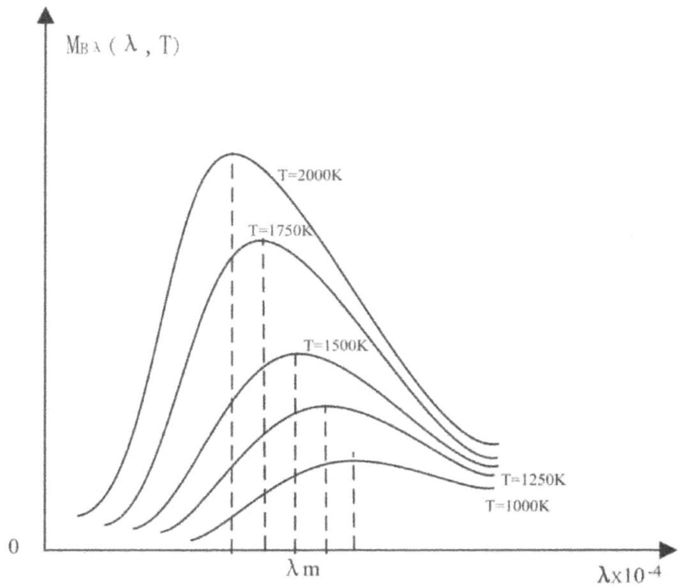

Figure (1-3) shows the absolute black-body energy spectrum curve they plotted.

1.2 Stefan–Boltzmann Law and Wien's Displacement Law

The work of O.R. Lummer and E. Pringsheim greatly advanced the research and exploration of thermal radiation theory, leading to corresponding theoretical achievements. Two important theoretical results emerged during this period: the Stefan–Boltzmann Law and Wien's Displacement Law.

1) Stefan (J. Stefan) – Boltzmann (L. Boltzmann) Law

In 1879, J. Stefan, after analyzing the absolute black body spectrum, deduced from experiments that the radiative ability of a black-body $R(T)$ is proportional to the fourth power of its thermodynamic temperature T, i.e.,

$$R_B(T) = R_{B\lambda}(\lambda,T) = \int_0^\infty r_{B\lambda}(\lambda,T)d\lambda = \sigma T^4 \qquad (1\text{-}12)$$

In this equation, σ is a universal constant, also known as the Stefan constant, with units of

$$\sigma = 5.67\times10^{-12} W/(cm^2.K^4) \qquad (1\text{-}13)$$

The experimental law summarized by Stefan was theoretically proven by Austrian physicist L. Boltzmann in 1884. This quantitative relationship is called the Stefan–Boltzmann Law. This law is commonly used in thermodynamics and astrophysics, for example, to estimate the radiative power of a star's surface.

2) Theoretically, W. Wien was the first to explore the mathematical form of the black body radiation energy spectrum. Using the similarity between the black-body radiation energy spectrum and the Maxwellian distribution of gas molecular velocities, Wien assumed that the distribution of thermal radiation according to wavelength followed a similar statistical law, leading to a theoretical formula:

$$R_{B\lambda}(\lambda,T) = B\lambda^{-5}e^{-A/\lambda T} \qquad (1\text{-}14)$$

In this equation, e is the base of the natural logarithm, and A and B are universal constants determined by experiments. Wien's work was highly creative; the conclusion he first derived—that the black body radiation energy $R_{B\lambda}(\lambda,T)$ is proportional to a negative exponential form of e and λ, which are related to T—was particularly remarkable. This conclusion ultimately played a pioneering role in the later discovery of the correct black body radiation law by Planck.

From Figure (1-3), we can see that the peak of each curve corresponds to the wavelength λ_m at which the radiation energy is maximized at that temperature, and at this point, the derivative of the energy spectrum function with respect to wavelength λ is zero. Differentiating equation (1-14), we get:

$$R'_{B\lambda}(\lambda,T) = B(\lambda^{-5}e^{-A/\lambda T})'$$

$$= B\left(-5\lambda^{-6}e^{-A/\lambda T} + \lambda^{-7}e^{-A/\lambda T}(A/T)\right) \qquad (1\text{-}15)$$

Setting $R'_{B\lambda}(\lambda,T) = 0$, we obtain:

$$\lambda_m T = \frac{A}{5} = const \qquad (1\text{-}16)$$

Equation (1-16) shows that the wavelength $R_{B\lambda}(\lambda,T)$ corresponding to the maximum value of the energy distribution function λ_m is inversely proportional to temperature T. $\lambda_m T$ is called the Wien constant. Now, the Wien constant can be accurately determined using the Planck formula, and experimental measurements give

$$\lambda_m T = 0.2897 cm.K \qquad (1\text{-}17)$$

Equations (1-14) and (1-17) are known as Wien's Displacement Law. Wien's Displacement Law forms the theoretical basis for high-temperature optical measurement techniques.

1.3 Rayleigh–Jeans Formula

After the release of the black body energy spectrum experimental curve by O.R. Lummer and E. Pringsheim, British physicist L. Rayleigh also engaged in the study of black body radiation laws. Based on classical electromagnetic theory and the equipartition theorem of energy, Rayleigh believed that the energy density in the cavity should be equal to the product of the electromagnetic state density in the cavity and the average energy of the standing waves. Since the electromagnetic state density in the cavity is the foundation for calculating the black body radiation spectrum, it is necessary to introduce it in detail here.

1) Electromagnetic State Density in the Cavity

Since the cavity is in a thermal equilibrium state, its radiation spectrum consists of standard spectral lines $u_T(\lambda)$ or $u_T(\nu)$. Any object, even a macroscopic object composed of single atoms, simultaneously excites electromagnetic radiation at different frequencies, indicating that the energy distribution state of the microscopic components of a macroscopic object is very complex, but it follows the Boltzmann energy distribution law. After excitation and absorption, the energy exchange between oscillators and the radiation field reaches thermal equilibrium, and there must be:

$$u_T(\nu) = g(\nu)\bar{\varepsilon}(\nu, T) \tag{1-18}$$

In this equation, $u_T(\nu)$ represents the energy spectral density in the cavity at temperature T, $g(\nu)$ is the number of standing waves in the frequency range from ν to $\nu + d\nu$ per unit volume, and $\bar{\varepsilon}(\nu, T)$ represents the average energy of standing waves in the frequency range from T to ν in the radiation field at temperature $\nu + d\nu$. Equation (1-18) means that the energy density in the cavity's radiation field equals the product of the state density and the average energy of the wave packet.

We know that for a one-dimensional linear wave, the necessary condition for forming a stable standing wave in the interval of length L is:

$$L = n\frac{\lambda}{2} \ (n = 1, 2, \ldots, n, \ldots) \tag{1-19}$$

where n is a natural number. Alternatively, k using the relationship between the wave vector λ and wavelength $(k = 2\pi/\lambda)$, it can also be written as:

$$k = n\frac{\pi}{L} \ (n = 1, 2, \ldots, n, \ldots) \tag{1-20}$$

The condition for the formation of a standing wave for radiation of wavelength L propagating along the azimuthal angle $\theta_x, \theta_y, \theta_z$ within a cube of side length λ is:

$$\begin{cases} L\cos\theta_x = n_1 \lambda/2 \\ L\cos\theta_y = n_2 \lambda/2 \\ L\cos\theta_z = n_3 \lambda/2 \end{cases} \tag{1-21}$$

or

$$\begin{cases} k_1 = k\cos\theta_x = n_1 \pi/L \\ k_2 = k\cos\theta_y = n_2 \pi/L \\ k_3 = k\cos\theta_z = n_3 \pi/L \end{cases} \tag{1-22}$$

where n_1, n_2, n_3 is a positive integer, and each set of (n_1, n_2, n_3) corresponds to a specific standing wave mode. $\cos\theta_x$, $\cos\theta_y$, $\cos\theta_z$ represents the direction cosine of the wave vector k, and since the sum of the squares of the direction cosines equals 1, from equation (1-22), we obtain:

$$k^2 = k_1^2 + k_2^2 + k_3^2 = \left(\frac{\pi}{L}\right)^2 \left(n_1^2 + n_2^2 + n_3^2\right) \tag{1-23}$$

Equation (1-23) can be understood as follows: we construct a wave vector space with k_1, k_2, k_3 as the axis, so that any wave vector corresponds to a coordinate point in the wave vector space. Using π/L as intervals, three mutually orthogonal coordinate planes divide the wave vector space into cubes with a side length of π/L, each with a volume of $(\pi/L)^3$. These cubes are called reciprocal cells, and they represent a possible standing wave mode. Since electromagnetic waves are transverse waves, there are two orthogonal polarization states, which correspond to two independent degrees of freedom of electromagnetic radiation. Therefore, the number of standing wave modes for wave vectors between 0 and k, denoted as $N(k)$, is equal to the number of reciprocal cells contained in the sphere with a radius of k in phase space, as shown in Figure (1-4).

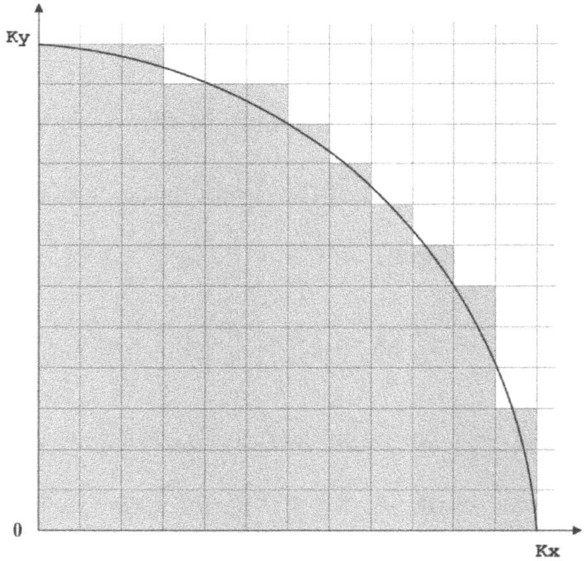

Figure (1-4) Schematic Diagram of Cavity Radiation State Density Modes.

Because n_1, n_2, n_3 are natural numbers, we take values in the first quadrant of phase space, and noting that the degrees of freedom of electromagnetic radiation are 2, the number of standing wave modes from 0 to k is:

$$N(k) = \frac{2}{8}\frac{4\pi k^3}{3}\left(\frac{L}{\pi}\right)^3 = \frac{k^3 L^3}{3\pi^2} \tag{1-24}$$

Since $k = 2\pi\nu/c$, equation (1-24) expressed in terms of frequency becomes:

$$N(\nu) = \frac{8\pi\nu^3 L^3}{3c^3} \tag{1-25}$$

Thus, the number of electromagnetic radiation modes in a unit volume within the frequency interval of ν is:

$$g(\nu) = \frac{1}{L^3}\frac{dN(\nu)}{d\nu} = \frac{8\pi\nu^2}{c^3} \tag{1-26}$$

With the cavity radiation field state density $g(\nu)$, the energy spectrum density of black-body radiation is:

$$u_T(\nu) = g(\nu)\bar{\varepsilon}(\nu, T) = \frac{8\pi\nu^2}{c^3}\bar{\varepsilon}(\nu, T) \tag{1-27}$$

2) Average energy of microscopic oscillators: From equation (1-27), it is known that the energy spectrum density of black-body radiation is determined by the product of the state density $g(\nu)$ and the average energy $\bar{\varepsilon}(\nu, T)$ of microscopic oscillators. Now

that the electromagnetic radiation standing wave spectrum density has been accurately given by classical electromagnetic theory in equation (1-26), once we calculate the average energy of the microscopic oscillators, we can obtain the energy spectrum density of blackbody radiation $u_T(\nu)$. Therefore, how to calculate the average energy of microscopic oscillators $\overline{\varepsilon}(\nu,T)$ becomes the key issue.

If we assume that the energy distribution of microscopic oscillators is continuous and follows the Boltzmann energy distribution law, the average energy of the oscillators $\overline{\varepsilon}(\nu,T)$ is given by the following formula:

$$\overline{\varepsilon}(\nu,T)=\frac{\int_0^a \varepsilon e^{-\varepsilon/kT}d\varepsilon}{\int_0^a e^{-\varepsilon/kT}d\varepsilon}=\left.\frac{(kT)^2(-\varepsilon/kT-1)e^{-\varepsilon/kT}}{-kTe^{-\varepsilon/kT}}\right|_0^a$$

$$=kT\left.\left(\frac{\varepsilon}{kT}+1\right)\right|_0^a \qquad (1\text{-}28)$$

where k is the Boltzmann constant ($k=1.38\times10^{-23}\ J/K$), T is the thermodynamic temperature of the cavity radiation field, and a is a sufficiently large natural number. It is especially important to note that 10^{26} should be understood as a sufficiently large natural constant, not as an infinitely large quantity. This is because, to date, we have not detected radiation with a frequency exceeding $\varepsilon\rightarrow0$

Hertz, and no oscillators with infinite energy exist in nature. This formula holds only when $\bar{\varepsilon}(\nu,T) \rightarrow kT$, and when $\varepsilon \rightarrow a$ a is sufficiently large, $\bar{\varepsilon}(\nu,T) \rightarrow \varepsilon$. This explains why the Rayleigh–Jeans formula agrees with experimental data in the long-wavelength, low-frequency range, but deviates significantly in the short-wavelength, high-frequency range.

According to modern quantum theory, the average energy of material oscillators must depend on frequency, and the average energy of oscillators at different frequencies cannot be the same universal constant. In fact, we know $\varepsilon_i = h\nu_i$.

On the other hand, any physical substance, as long as its temperature is not absolute zero, radiates electromagnetic radiation at different frequencies ν_i ($i = 1,2,..., n,...$) to the surrounding space. If we call ν_i the fundamental frequency of the i-th mode, there will be high-frequency oscillators with a frequency of $\nu_{mi} = m\nu_i$ $(m = 1,2,...)$ for the same fundamental frequency. Thus, when we divide the energy interval by $d\varepsilon$, oscillators with a frequency of $\nu_{mi} = \nu_{nj}$ fall into the same interval. However, they have different fundamental frequencies and belong to different state density functions $g(\nu_i)$ and $g(\nu_j)$, which is why equation (1-27) fails to accurately reveal the blackbody radiation law. It is significant that the Rayleigh–Jeans formula agrees with experimental results in the low-frequency range, while the Wien formula agrees with experimental results in the high-frequency range. These two complementary results significantly contributed to Planck's discovery of the correct black-body radiation law.

S.2 Planck's Black-body Radiation Law

2.1 Planck's Early Work

After the publication of the W. Wien distribution law in 1893, it attracted widespread attention in the field of theoretical physics, and theorists sought to integrate the Wien distribution law into the thermodynamic theory. Theoretical physicist Planck believed that the derivation of the Wien law made too many assumptions and that its conclusions were not convincing. Starting in 1897, Planck devoted himself to the theoretical study of black-body radiation, applying electromagnetic theory to thermal radiation. Starting from the energy exchange and interaction between oscillators and the radiation field, and using the calculation of the Boltzmann entropy, he derived the Wien distribution law again, thus giving it universal significance.

However, the Wien distribution law showed systematic deviations from the Lummer–Pringsheim experimental curve in the long-wavelength range, indicating that the Wien distribution law was not yet complete. At this time, British physicists Rayleigh and Jeans published their research findings. The Rayleigh–Jeans blackbody radiation energy spectrum formula, although deviating significantly from experimental data in the short-wavelength range, agreed well with experimental data in the long-wavelength range. The deviations between the Wien and Rayleigh–Jeans theoretical curves and the experimental curves are shown in Figure (2-1).

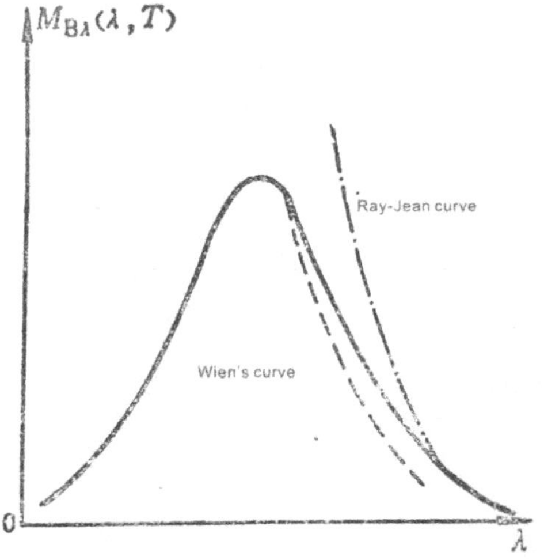

Figure (2-1) Deviations between the Wien and Rayleigh-Jeans theoretical curves and experimental data.

This situation led Planck to seek a unified formula that would asymptotically approach the Wien formula in the short-wavelength range and the Rayleigh-Jeans formula in the long-wavelength range. After a period of intensive derivation, Planck, in collaboration with experimental physicist Rubens, successfully derived his blackbody radiation energy spectrum formula:

$$u_T(\lambda) = A\lambda^{-5} \frac{1}{e^{B/\lambda T} - 1} \tag{2-1}$$

Compared to the Wien formula, this formula only differs by an additional (-1) after the exponential function, yet miraculously matches the experimental data very closely. This formula is known as Planck's black-body radiation formula.

2.2 The Final Form of Planck's Law

Although equation (2-1) agrees highly with experimental data, like the Wien formula, it still contains two constants, A and B, that require experimental determination. Moreover, the physical meaning of equation (2-1) was not yet clear. Planck was not satisfied with the fact that equation (2-1) merely aligned with experimental facts; he aimed to further reveal the physical significance of the black-body radiation laws inherent in the formula. At that time, the Boltzmann energy distribution law had already been established and was widely accepted in the theoretical community. Although the Boltzmann energy distribution law was derived from rigorous statistical reasoning based on ideal gases, there were indications that it also applied to solid-state materials. On the other hand, it was known that any material at a non-zero temperature radiates electromagnetic radiation at multiple frequencies simultaneously. These frequencies were called fundamental frequencies or intrinsic frequencies, denoted as ν_i $(i = 1,2,..., n,...)$. The energy of oscillators at different frequencies also varies, and according to the Boltzmann energy distribution law, the distribution probabilities of oscillators at different frequencies differ. However, if the energy of oscillators were continuously distributed, the average energy $\overline{\varepsilon}(\nu,T)$ of an oscillator would inevitably be calculated according to equation (1-28). But this did not match experimental results.

Planck believed that even for oscillators with the same fundamental frequency ν_i, their energies were not necessarily the same, and situations like ν_{ik} $(k = 1,2,..., m,...)$ also existed. However, if energy were continuously distributed, even for an oscillator with frequency ν_i, the average energy $\overline{\varepsilon}_i(\nu,T)$ would still be the result from equation (1-28). Planck was astonished to discover that if

$$\nu_{ik} = k\nu_i \ (k = 1,2,3,...) \tag{2-2}$$

held (where k is a natural number), then the method for calculating the average energy via level integration could be replaced by summing a series. Furthermore, the average energy calculated using the series summation method precisely took the form of equation (2-1). After a period of difficult and careful thinking, Planck finally proposed his quantum theory of energy, breaking through the classical concept of continuous energy distribution and opening a new chapter in quantum physics. In this sense, Planck's work cannot be praised highly enough.

Planck believed that the energy spectrum density of the radiation field $u(v,T)$ is determined by the state density of the radiation field and the average energy of the oscillators. However, in calculating the average energy of the oscillators, Planck introduced the revolutionary concept of energy quanta, founding the theory of energy quanta. According to this theory, the energy of an oscillator is discontinuous and has a minimal unit. For any oscillator with frequency v, the smallest unit of energy is $\varepsilon = hv$. Furthermore, the energy of an oscillator with frequency ν can only take one of the discrete energy states such as ε, 2ε, 3ε, $\varepsilon_i = i\varepsilon = ihv$, $(i = 1,2,...)$. Here, Planck's assumption of equally spaced energy levels for oscillators was later rigorously proven by quantum mechanics. According to the Boltzmann energy distribution law, the normalized distribution law for oscillators with frequency v and energy levels ε_0, ε_1, ε_2, ε_3, ... in a cavity material is given by:

$$B(\varepsilon_n,T)d\varepsilon = De^{-\varepsilon_n/kT}d\varepsilon \ (n = 1,2,...,i,...) \qquad \text{(2-3)}$$

where k is the Boltzmann constant, and D is the normalization constant

$$D = \frac{1}{\sum_{n=0}^{\infty} e^{-\varepsilon_n/kT} \Delta\varepsilon} \tag{2-4}$$

Since the energy levels are discrete, the average energy of the oscillators must be directly calculated by summation. Noting that $\varepsilon = hv$, the average energy is:

$$\bar{\varepsilon} = \frac{\sum_{n=0}^{\infty} \varepsilon_n e^{-\varepsilon_n/kT} \Delta\varepsilon}{\sum_{n=0}^{\infty} e^{-\varepsilon_n/kT} \Delta\varepsilon} = \frac{\sum_{n=0}^{\infty} nhve^{-nhv/kT} \Delta\varepsilon}{\sum_{n=0}^{\infty} e^{-nhv/kT} \Delta\varepsilon}$$

$$= (-1)\frac{\partial}{\partial(1/kT)} \ln\left(\sum_{n=0}^{\infty} e^{-nhv/kT}\right)$$

$$= (-1)\frac{\partial}{\partial(1/kT)} \ln\left(1 - e^{-hv/kT}\right)^{-1}$$

$$= \frac{hve^{-hv/kT}}{1 - e^{-hv/kT}} = \frac{hv}{e^{hv/kT} - 1} \tag{2-5}$$

If the state density in the $v + dv$ frequency range is expressed by $g(v)dv$, and using the Rayleigh-Jeans calculation result from equation (1-28), the black-body radiation energy spectrum density is expressed by $u(v,T)$. Therefore, Planck's blackbody radiation formula is:

$$u(v,T) = g(v)\bar{\varepsilon}dv$$

$$= \frac{8\pi v^2}{c^3} \frac{hv}{e^{hv/kT} - 1} dv \tag{2-6}$$

Before Planck, Rayleigh had computed the electromagnetic standing wave state density in a cavity based on the assumption of electromagnetic media in vacuum and gave the black-body energy spectrum formula:

$$u(v,T)=\frac{8\pi v^2}{c^3}kT$$

The formula agrees with experimental values in the long-wavelength range but deviates significantly in the short-wavelength range. It should be said that the Rayleigh formula is the inevitable conclusion of classical electromagnetic theory and the energy equipartition law. So, where does the Rayleigh-Jeans formula go wrong? In retrospect, it is clear that one cannot directly use state density as the basis for calculating the number of standing waves, as this would inevitably lead to the conclusion that the higher the frequency, the more standing waves there are, causing the radiation energy to tend to infinity, known as the "ultraviolet catastrophe." In fact, the number of standing waves in a cavity is determined by two independent factors: it is not only related to state density but also to the distribution probability of oscillators with frequency v in the cavity. This conclusion is not hard to understand, just like how the speed of light, c, is the ultimate speed limit, nature also imposes a limit on the wavelength, λ_m, and a limit on the frequency, v_m.

Experimental evidence shows that, to date, even the strongest cosmic rays detected have frequencies not exceeding the 10^{26} Hertz range. Oscillators with frequencies higher than the limit frequency ν_m cannot exist in any known material. However, the state density given by the Rayleigh-Jeans formula is surprisingly high. Moreover, when calculating the average energy of standing waves, it relies on the classical theory of continuous energy distribution.

2.3 Verification of Planck's Formula

After the publication of Planck's blackbody radiation formula, it not only received support from experimental data but was also widely recognized in the theoretical community. This is demonstrated by the fact that starting from Planck's formula, one can naturally derive the Wien formula, Rayleigh-Jeans formula, Stefan-Boltzmann law, and Wien's displacement law, among other earlier theoretical results on blackbody radiation.

1) Derivation of the Wien Formula

The Wien formula agrees with experimental data in the high-frequency range. The wavelength form of Planck's formula (2-6) is:

$$u(\lambda,T)=\frac{8\pi hc}{\lambda^5}\frac{1}{e^{hc/kT\lambda}-1} \tag{2-7}$$

Let $\lambda\to\varepsilon$, and let ε be a sufficiently small positive real number, such as ε, which is the wavelength of ultraviolet light smaller than 10^{-8} cm. At this point, $(e^{hc/kT\lambda}-1)\to e^{hc/kT\lambda}$, so we have:

$$u(\lambda,T)_{\lambda\to\varepsilon}=8\pi hc\lambda^{-5}e^{-hc/kT\lambda} \tag{2-8}$$

Equation (2-8) is the formal form of the Wien distribution. Compared with equation (1-14), it contains $B=8\pi hc$, $A=hc/k$.

2) Derivation of the Rayleigh-Jeans Formula

The Rayleigh-Jeans formula agrees with experimental data in the low-frequency, long-wavelength range. Using Planck's formula in equation (2-5), when the frequency ν is sufficiently small, for example, when ν is smaller than the frequency of infrared light, 10^{10} Hz, $e^{h\nu/kT}\to(1+h\nu/kT)$ we have:

$$u(\nu,T)_{\nu\to a}=\frac{8\pi\nu^2}{c^3}\frac{h\nu}{(1+h\nu/kT)-1}$$

$$=\frac{8\pi\nu^2}{c^3}kT \tag{2-9}$$

In this equation, a is less than the frequency of infrared light, k is the Boltzmann constant, and T is the system's thermodynamic temperature.

3) Derivation of the Stefan-Boltzmann Law

Using Planck's formula, we can derive the Stefan-Boltzmann law. At this point, we have:

$$R(T)=\int_0^{\infty} r(\nu,T)d\nu=\frac{c}{4}\int_0^{\infty} u(\nu,T)d\nu$$

$$=\frac{2\pi h}{c^2}\int_0^{\infty}\frac{\nu^3}{e^{h\nu/kT}-1}d\nu \qquad (2\text{-}10)$$

Let $x=h\nu/kT$, then $d\nu=(kT/h)dx$, and after substituting into the above equation:

$$R(T)=\frac{2\pi k^4T^4}{c^2h^3}\int_0^{\infty}\frac{x^3}{e^x-1}dx$$

$$=\frac{2\pi k^4T^4}{c^2h^3}\Gamma(4)\zeta(4)=\frac{2\pi k^4}{c^2h^3}\frac{\pi^4}{15}T^4 \qquad (2\text{-}11)$$

Let

$$\sigma=\frac{2\pi^5k^4}{15c^2h^3} \qquad (2\text{-}12)$$

Then we have $R(T)=\sigma T^4$, which gives the Stefan-Boltzmann law.

4) Derivation of the Wien Displacement Law

The Wien Displacement Law has the form: $T\lambda_{\max}=b$. We differentiate Planck's formula in the wavelength form and set its derivative equal to zero, i.e.,

$$d\left(\frac{8\pi hc}{\lambda^5}\frac{1}{e^{hc/kT\lambda}-1}\right)=0 \qquad (2\text{-}13)$$

After simplification, we have:

$$\frac{hc}{kT\lambda}e^{hc/kT\lambda}-5e^{hc/kT\lambda}+5=0 \tag{2-14}$$

Noting that $(h\nu/kT\lambda) > 0$, we let $(hc/kT\lambda)=x$, and the above expression simplifies to:

$$xe^{x}-5x+5=0 \tag{2-15}$$

The equation has a non-zero solution, which can be obtained by approximation as $x=4.965 \quad (hc/kT\lambda)=4.965$. That is,

$$T\lambda_{\max}=hc/k\times 4.965=0.2041(hc/k) \tag{2-16}$$

This equation shows that the product of the peak value of the cavity radiation energy spectrum density and temperature is a universal constant. Thus, the Wien Displacement Law is proven. In equation (2-16), $\lambda_{\max}$ is the wavelength of radiation corresponding to the maximum energy spectrum density T and monochromatic emissivity $u(\lambda,T)$ when the radiation field temperature is $r(\lambda,T)$.

Before we end this section, it is still meaningful to review the thought process through which Planck introduced the concept of energy quanta. As mentioned earlier, Planck, with the help of the Wien and Rayleigh-Jeans formulas, quickly derived the black-body radiation energy spectrum formula, which is equation (2-1). However, as a theoretical physicist, he was not satisfied with this interpolated formula. He later summarized: "Even if this new radiation formula is proven to be absolutely accurate, if it is merely an interpolated formula arrived at by chance, its value would be limited." Therefore, after proposing this formula, I committed myself to exploring the true physical meaning behind it. Initially, Planck, using general thermodynamic theory and classical electromagnetic theory, studied blackbody radiation but failed to derive the experimentally validated formula. However, when Planck accepted Boltzmann's statistical thermodynamics, a major breakthrough in theoretical research occurred.

Planck's Discovery Based on Boltzmann's Statistical Thermodynamics

According to Boltzmann's statistical thermodynamics method, the total energy of the system must first be divided into several small portions, and then each portion is assigned to the system's oscillators. Let the total energy of the system be E. Assuming the smallest energy unit is ε, and the number of divisions is m, then we have:

$$E = m\varepsilon . \tag{2-17}$$

If the number of oscillators in the system is N, and assuming each oscillator can receive energy units ε in varying proportions, if the total number of ways to distribute the energy across the oscillators is W, also known as the configuration number, then:

$$W = \frac{(N+m-1)!}{(N-1)!m!} \tag{2-18}$$

Using Stirling's approximation, $\ln x! = x\ln x - x$, from the above equation, we get:

$$\ln W = (N+m-1)\ln(N+m-1)-(N+m-1)$$

$$-(N-1)\ln(N-1)+(N-1)-m\ln m+m$$

$$=(N+m-1)\ln(N+m-1)-(N-1)\ln(N-1)-m\ln m \qquad (2\text{-}19)$$

Noting that N, m is much greater than 1, the equation can be written as:

$$\ln W = (N+m)\ln(N+m)-N\ln N-m\ln m \qquad (2\text{-}20)$$

According to Boltzmann's entropy theory, the entropy S_N of the oscillator system is proportional to $\ln W$. Specifically,

$$S_n = k\ln W \qquad (2\text{-}21)$$

where k is the Boltzmann constant. If the average energy of an oscillator is $\bar{\varepsilon}$, then

$$\bar{\varepsilon} = \frac{E}{N} = \frac{m\varepsilon}{N}, \quad N = \frac{m\varepsilon}{\bar{\varepsilon}} \qquad (2\text{-}22)$$

On the other hand, the entropy S_N of a system of N oscillators is N times the entropy S of a single oscillator, i.e., $S_N = NS$. Therefore, the average entropy of a single oscillator is:

$$S = \frac{S_N}{N} = \frac{k}{N}\ln W \qquad (2\text{-}23)$$

Substituting equation (2-22) into this, and noting the properties of the logarithmic function, we simplify to:

$$S = k\left[\left(1+\frac{\bar{\varepsilon}}{\varepsilon}\right)\ln\left(1+\frac{\bar{\varepsilon}}{\varepsilon}\right)-\frac{\bar{\varepsilon}}{\varepsilon}\ln\frac{\bar{\varepsilon}}{\varepsilon}\right] \tag{2-24}$$

Differentiating both sides of the above equation gives:

$$dS = k\left[\frac{1}{\varepsilon}\ln\left(1+\frac{\bar{\varepsilon}}{\varepsilon}\right)+\left(1+\frac{\bar{\varepsilon}}{\varepsilon}\right)\left(1+\frac{\bar{\varepsilon}}{\varepsilon}\right)^{-1}-\frac{1}{\varepsilon}\ln\frac{\bar{\varepsilon}}{\varepsilon}-\frac{\bar{\varepsilon}}{\varepsilon}\left(\frac{\bar{\varepsilon}}{\varepsilon}\right)^{-1}\right]d\bar{\varepsilon}$$

$$=\frac{k}{\varepsilon}\left[\ln\left(1+\frac{\bar{\varepsilon}}{\varepsilon}\right)-\ln\frac{\bar{\varepsilon}}{\varepsilon}\right]d\bar{\varepsilon} \tag{2-25}$$

According to the second law of thermodynamics, $d\bar{\varepsilon} = TdS$, we can solve for:

$$\frac{1}{T}=\frac{dS}{d\bar{\varepsilon}}=\frac{k}{\varepsilon}\left[\ln\left(1+\frac{\bar{\varepsilon}}{\varepsilon}\right)-\ln\frac{\bar{\varepsilon}}{\varepsilon}\right] \tag{2-26}$$

Thus, Planck derived the average energy of the oscillators:

$$\bar{\varepsilon}=\frac{\varepsilon}{e^{\varepsilon/kT}-1} \tag{2-27}$$

At this point, Planck realized that the average energy of the oscillators, calculated through Boltzmann's entropy, had the negative exponential form of the Wien formula. The Wien formula shows that the energy spectrum density is related to the frequency and wavelength of the oscillators. Given this, Planck believed that the entropy of an oscillator should be a function of the average energy and frequency or wavelength. Ultimately, Planck concluded that the energy unit envisioned by Boltzmann takes the following form:

$$\varepsilon = h\nu \tag{2-28}$$

In fact, oscillators have multiple frequency modes. For each oscillator with an inherent frequency of ν_i, the minimum energy unit is:

$$\varepsilon_i = h\nu_i \ (i = 1,2,\dots, i,\dots) \tag{2-29}$$

The average energy is:

$$\bar{\varepsilon}_i = \frac{\varepsilon_i}{e^{\varepsilon_i/kT} - 1} = \frac{h\nu_i}{e^{h\nu_i/kT} - 1} \ (i = 1,2,\dots, i,\dots) \tag{2-30}$$

Multiplying equation (2-30) by the state density $g(\nu)$, i.e., equation (2-7), gives the clear physical meaning to the black-body radiation formula. The constant h in the equation is known as Planck's constant. The value calculated by Planck was: $h = 6.65\times 10^{-34}\ (J.s)$. The current precise value is:

$$h = 6.62607\times 10^{-34}\ (J.s)\,. \tag{2-31}$$

On the other hand, Boltzmann had the idea of infinitely subdividing energy, but still considered energy to have continuity. He did not provide criteria for the subdivision. Planck recognized that energy has a minimum unit. Inspired by the Wien formula, he hypothesized that for oscillators with a frequency of ν, the minimum energy unit is $h\nu$, and the oscillators can only exist in one of a series of energy states, namely $h\nu, 2h\nu, \dots, ih\nu, \dots$. The energy level spacing is $h\nu$.

Planck's concept of energy quantization was groundbreaking. Under the influence of this idea, in 1905, Einstein (A. Einstein) proposed the concept of the photon. He extended energy quantization from oscillators to photons, proposing that the energy of a photon (wave packet) is proportional to the frequency of the photon, i.e., the energy of the photon $\varepsilon = h\nu$, with Planck's constant h as the proportionality coefficient. This successfully explained the photoelectric effect experiment. In 1913, Bohr (N. Bohr) proposed the theory of quantization of hydrogen atom energy levels, solving the problem of the hydrogen atom spectrum. In 1923, de Broglie (L. de Broglie) proposed the hypothesis of wave-particle duality of electrons. The energy and momentum are:

$$E = h\nu, \qquad P = \frac{h}{\lambda} \tag{2-32}$$

where ν and λ are the frequency and wavelength of the electron wave packet, respectively. Subsequently, Heisenberg (W. Heisenberg) established matrix mechanics in 1925, and in 1926, Schrödinger (E. Schrödinger) founded wave mechanics, opening a new chapter in quantum physics.

Planck's constant h is an important constant in the field of micro physics, as the energy, momentum, angular momentum, and other properties of microscopic particles are all related to this constant. However, our research has found that Planck's constant h

is not a fundamental constant but is a composite quantity composed of a set of characteristic quantities of quantum space. In other words, Planck's constant h has an analytical expression. The analytical expression for h we propose is:

$$h = 2\pi^2 \rho \varphi_0^2 c \tag{2-33}$$

where ρ is the mass density of quantum space, φ_0 is the minimum limit amplitude of quantum space, and c is the rate at which radiation propagates in free quantum space, i.e., the speed of light. The values for ρ and φ_0 are both quantitatively determined. However, we believe that this discussion pertains to the fundamental laws of thermal radiation, and it is not appropriate to expand on the analytical expression of Planck's constant here. We will discuss this sensitive topic in the appendix, and even interested readers are advised to read the appendix after reading the main text.

S.3 Laws of Energy Exchange Between Oscillators and the Radiation Field

From galaxies and planets to the microscopic systems of matter, energy exchange is continuous and ubiquitous. Radiation is the most universal and fundamental form of energy exchange between all things in the universe. Therefore, revealing the laws of energy exchange between microscopic material oscillators and the radiation field is of great significance for both thermodynamics and cosmology. This is the core topic of this book. Energy exchange between microscopic oscillators and the radiation field involves two aspects: one is the radiation emitted by the oscillators, and the other is the radiation absorbed by the oscillators. We will first discuss the laws of radiation emitted by oscillators.

3.1 Laws of Radiation Emission by Oscillators

To find the law of radiation emission by oscillators, we need to recalculate the number of wave packets in the frequency range $v+dv$. For a cavity radiation field at a temperature of T, we express the probability of an oscillator with frequency $p_i(hv,T)$

and energy v emitting radiation, i.e., the probability of the oscillator emitting photons. We assume that oscillators with frequencies of ihv and energy levels of $0hv$, $1hv$, $2hv$, ..., ihv, ..., transition between the same fundamental frequency energy levels, emitting photons of frequency v. However, the probability of oscillators emitting radiation at different energy levels varies. High-energy oscillators are more likely to transition and emit radiation than low-energy oscillators. Let $k_i(hv,T)$ represent the probability (rate) of an oscillator with frequency v and energy ihv emitting radiation per unit time. The probability of radiation emission by an oscillator must be proportional to the oscillator's energy level. That is, the probability of an oscillator emitting radiation with energy levels of` hv, $2hv$, $3hv$, ... , is:

$$k_1(hv,T):k_2(hv,T):k_3(hv,T):...=1:2:3:... \qquad (3\text{-}1)$$

For oscillators with frequency v and energy ihv, the distribution function is given by equation (2-3). The probability of radiation emission by these oscillators is determined by both the oscillator's energy level state and the energy distribution probability. That is, the probability of radiation emission by an oscillator at energy level i is:

$$p_i(hv,T)=k_ie^{-hv/kT}=ie^{-hv/kT} \qquad (3\text{-}2)$$

Combining equation (3-1) and equation (2-3), the probability of radiation emission by oscillators with different energy levels at frequency v in the cavity is given by:

$$p_1(hv,T):p_2(hv,T):...=e^{-hv/kT}:2e^{-2hv/kt}:... \tag{3-3}$$

Thus, the average probability of radiation emission by oscillators at frequency v in the cavity is:

$$f(v,T)=\frac{\sum_{i=0}^{\infty} ie^{-ihv/kT}}{\sum_{i=0}^{\infty} e^{-ihv/kT}}=\frac{1}{e^{hv/kT}-1} \tag{3-4}$$

If we express $n(v,T)$ as the number density of standing waves in the frequency range v per unit frequency interval in the cavity, it is clear that the number of standing waves in the cavity is related to both the state density and the average emission probability. The product of these two determines the number of standing waves. Therefore, in the frequency range v, the number density of standing waves is:

$$n(v,T)=g(v,T)f(v,T)$$

$$=\frac{8\pi v^2}{c^3}\frac{1}{e^{hv/kT}-1} \tag{3-5}$$

The energy of the standing waves is equal to the energy of the oscillations at the same wavelength. Multiplying equation (3-5) by the energy of a wave packet, hv, gives us Planck's formula (2-7). The fact that Planck's formula agrees with experimental data means that the number of standing waves represented by equation (3-5) is consistent with experimental results. Consequently, the radiation probability relationships revealed by equations (3-1) and (3-3), which are proportional to the oscillator's energy state, align with reality. Therefore, we say that Planck's formula not only reveals the laws of blackbody radiation but also includes the proportional relationship between the probability of radiation emission and the oscillator's energy state. Equations (2-5) and (3-4) indicate that the average emission probability of oscillators at frequency ν in the cavity is proportional to the average energy of those oscillators.

Based on this, the conclusion about the laws of radiation emission by oscillators is: The probability of radiation emission by material oscillators in a thermal radiation field is proportional to the oscillator's energy state. For each specific oscillator, the probability of radiation emission is proportional to its energy state. For a large number of oscillators in the cavity, the average probability of radiation emission is proportional to their average energy, with the average probability given by equation (3-4).

The law of radiation emission by oscillators is contained within Planck's blackbody radiation law. By changing our approach and calculating the average emission probability of the oscillators, we naturally obtain this law. However, the law of radiation absorption by oscillators is not as obvious.

3.2 Laws of Radiation Absorption by Oscillators

Energy exchange between oscillators and the radiation field includes both excitation and absorption. We have already discussed the law of radiation emission by oscillators, and now, based on the Boltzmann energy distribution law and the newly derived law of radiation emission by oscillators, we analyze the law of radiation absorption by oscillators. The energy and frequency distribution of material oscillators in a thermal equilibrium radiation field is shown in Table (1). In Table (1), the vertical axis represents the Boltzmann distribution of the oscillator's energy, while the horizontal axis represents the Planck distribution of the oscillator's frequency. N_i and n_i represent the corresponding numbers of oscillators. The horizontal line in the middle of the table represents energy levels, with energy levels being evenly spaced.

Boltamann distribution									
energy	distribution	n_0	n_1	n_2	.....	n_i	.....	n_m	
$n\varepsilon+d\varepsilon$	$B(\varepsilon_n)$								N_n
⋮	⋮								⋮
$i\varepsilon+d\varepsilon$	$B(\varepsilon_i)$								N_i
⋮	.....								⋮
$2\varepsilon+d\varepsilon$	$B(\varepsilon_2)$								N_2
$1\varepsilon+d\varepsilon$	$B(\varepsilon_1)$								N_1
$0\varepsilon+d\varepsilon$	$B(\varepsilon_0)$								N_0
		v_0+dv	v_1+dv	v_2+dv	.....	v_i+dv	.....	v_m+dv	frequency · planck distribution
		$P(v_0)$	$P(v_1)$	$P(v_2)$	......	$P(v_i)$	......	$P(v_m)$	distribution

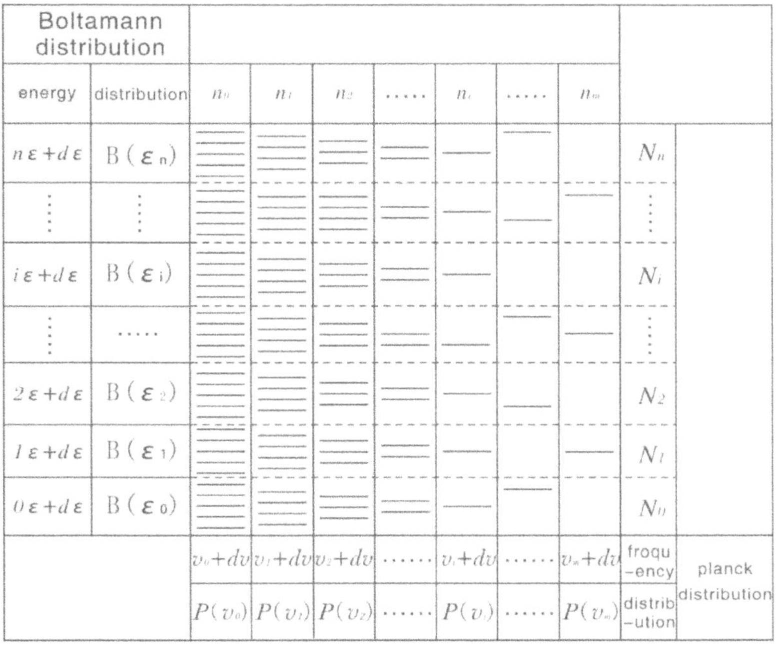

Figure (3-1) Schematic Diagram of Oscillator Energy Levels and Frequency Distribution

In Table (1), $B(v,T)$ represents the normalized Boltzmann energy distribution function, and $P(v,T)$ is the Planck frequency distribution formula.

Einstein (A. Einstein) distinguished the energy exchange in thermal radiation into three different processes: spontaneous radiation, stimulated radiation, and stimulated absorption. Spontaneous radiation is radiation emitted by oscillators due to spontaneous transitions, where the photons emitted by the oscillator have randomness in polarization state, phase, and propagation direction. In contrast, stimulated radiation is radiation emitted by oscillators when they are stimulated by incident photons from the radiation field, maintaining the same frequency, phase, polarization state, and propagation direction as the incident photons. The radiation field inside the cavity is the result of both types of radiation.

The absorption process is the reverse: oscillators can only absorb photons that are resonant with their frequency. Since the state density in the cavity is discrete and photons are non-degenerate, oscillators can only absorb one photon of the same frequency at a time. This is the essential feature of an oscillator's photon absorption. The energy levels of harmonic oscillators are equidistant, and if the frequency is v_i, the oscillator transitions from the energy level n to the energy level m, crossing $(n-m)$ energy levels and exciting $(n-m)$ photons with a frequency of v_i. This is not the same as exciting a single photon with a frequency of $(n-m)v_i$, which differs from atomic energy level transitions. In the table, k_i $(i = 1,2,3...)$ represents the number of energy levels of oscillators at different frequencies within the same energy range. Low-frequency oscillators have a larger value of k, indicating dense energy levels, while high-frequency oscillators have a smaller value of k, indicating greater energy level spacing.

Cavity radiation is thermal equilibrium radiation. To find the law of oscillator absorption of radiation, we consider an extreme situation where the external work makes the cavity reach absolute zero, and the material oscillators with different frequencies, distributed according to the Boltzmann energy distribution law, emit all the corresponding frequency photons they carry within a period of time, returning to the ground state. Then, we let the cavity absorb photons again under the original environment, allowing the energy to rearrange from the ground state and eventually return to the previous state. According to the Boltzmann energy distribution law, even in this case, under constant macroscopic temperature, the energy distribution of the oscillators will still follow the previously established distribution.

According to quantum mechanics, ground-state oscillators have the lowest energy, and at this point, the oscillators do not emit photons. A ground-state oscillator with frequency v has an energy of $(E_0 = hv/2)$. Assuming that all the oscillators emit photons of various frequencies, the probability of the oscillator being in the ground state is 1 $(G_0 = 1)$. Here, we divide the energy of an oscillator with frequency v into several equal parts, each with a minimum energy unit of $\Delta\varepsilon = hv$. At this point, the energy level distribution normalization factor is:

$$D = \frac{1}{\sum_{i=0}^{\infty} e^{-ihv/kT} \Delta\varepsilon} = \frac{(1 - e^{-hv/kT})}{\Delta\varepsilon} \tag{3-6}$$

If the probability that the oscillator absorbs at least one photon with frequency v is G_1, then:

$$G_1 = 1 - De^{-0hv/kt}\Delta\varepsilon = 1 - D\Delta\varepsilon$$

$$= 1 - \left(1 - e^{-hv/kT}\right) = e^{-hv/kT} \tag{3-7}$$

Equation (3-7) represents the probability that the oscillator absorbs at least one photon, i.e., the probability that the oscillator is not in the ground state. The concept is clear: absorbing one photon is absorption, and absorbing i photons is also absorption. According to the characteristics of oscillator photon absorption, absorbing i photons always follows the absorption of one photon. Therefore, any oscillator with a level higher than the ground state is within this probability range, so the probability of a ground-state oscillator absorbing radiation is given by equation (3-7). The probability of an oscillator absorbing one photon is illustrated in Figure (3-2)(a).

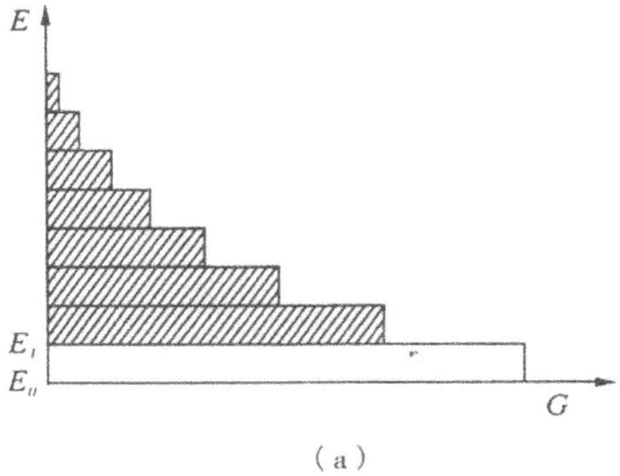

Figure (3-2)(a) Schematic Diagram of the Probability of an Oscillator Absorbing One Photon.

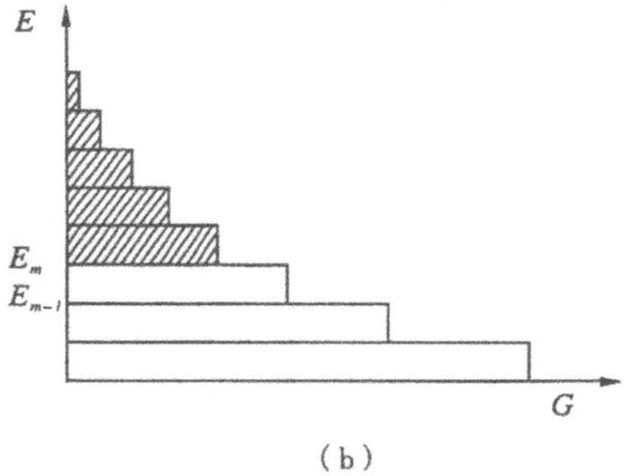

Figure (3-2)(b) Schematic Diagram of the Probability of an Oscillator Absorbing m Photons.

Similarly, for an oscillator in the E_1 state, the probability of absorbing radiation is:

$$
\begin{aligned}
G_2 &= 1 - D\left(1 + e^{-h\nu/kt}\right)\Delta\varepsilon \\
&= 1 - \left(1 - e^{-h\nu/kT}\right)\left(1 + e^{-h\nu/kT}\right) \\
&= e^{-2h\nu/kT} \qquad (3\text{-}8)
\end{aligned}
$$

The meaning of equation (3-8) is the probability that the oscillator absorbs at least two photons. Therefore, any oscillator with an energy level higher than E_1 is within this probability range. Similarly, we define G_2 as the probability of an oscillator in the energy level E_1 state absorbing radiation. Likewise, we define the probability of an oscillator in the energy level E_2 state absorbing radiation as G_3, then

$$G_3 = 1 - D\left(1 + e^{-hv/kT} + e^{-2hv/kT}\right)\Delta\varepsilon$$
$$= 1 - \left(1 - e^{-hv/kT}\right)\left(1 + e^{-hv/kT} + e^{-2hv/kt}\right)$$
$$= e^{-3hv/kT}$$

As shown in Figure (3-2)(b), in general, the probability that an oscillator in the energy level E_{m-1} state absorbs radiation is

$$G_m = 1 - D\sum_{i=0}^{m-1} e^{-ikv/kT}\Delta\varepsilon = 1 - \frac{\sum_{i=0}^{m-1} e^{-ihv/kT}\Delta\varepsilon}{\sum_{i=0}^{n\to\infty} e^{-ihv/kT}\Delta\varepsilon}$$

$$= 1 - \left(\frac{1 - e^{-mhv/kY}}{1 - e^{-hv/kT}} \Big/ \frac{1 - e^{-(n+1)hv/kT}}{1 - e^{-hv/kT}}\right)$$

$$= \frac{e^{-mhv/kT} - e^{-(n+1)hv/kT}}{1 - e^{-(n+1)hv/kT}} \tag{3-9}$$

In this equation, n represents the highest energy level of the oscillator. Noting that when n is sufficiently large, $e^{-(n+1)h\nu/kT} \to 0$, the equation (3-9) can be expressed as:

$$G_m = e^{-mh\nu/kT} \tag{3-10}$$

Equation (3-10) holds for oscillators at various frequencies inside the cavity. It shows that the probability of an oscillator absorbing radiation is proportional to the probability of its energy distribution. Since:

$$\begin{aligned} &G_0 + G_1 + G_2 + \ldots + G_m + \ldots \\ &= 1 + e^{-h\nu/kT} + e^{-2h\nu/kT} + \ldots + e^{-mh\nu/kT} + \ldots \\ &= \sum_{m=0}^{m\to\infty} e^{-mh\nu/kT} \end{aligned} \tag{3-11}$$

The normalization factor is still D. After normalization, the probability of an oscillator in the E_{m-1} energy state absorbing radiation is: $De^{-mh\nu/kT}$. However, for an oscillator in the E_{m-1} energy state, i.e., an oscillator that has absorbed $(m-1)$ photons, the probability of the oscillator in the E_{m-1} energy state absorbing radiation is precisely the probability of the oscillator absorbing m photons. Thus, the average probability of an oscillator absorbing radiation can be written as:

$$\overline{G} = D\sum_{m=0}^{m\to\infty} me^{-mh\nu/kT} = \frac{1}{e^{h\nu/kT} - 1} \tag{3-12}$$

Equation (3-12) is equivalent to equation (3-4), now representing the average probability of an oscillator absorbing radiation. Similarly, the probability of an oscillator absorbing radiation is not only related to the average absorption probability, but also to the state density of oscillators at that frequency inside the cavity. The product of the average absorption probability and the state density gives the probability of an oscillator absorbing radiation (photons). We know $\varepsilon = h\nu$. Multiplying by $h\nu$ gives us the Planck black-body radiation formula. This formula is now derived from the opposite process of absorption, thus providing a theoretical explanation for the energy exchange law in thermal equilibrium radiation.

Now, regarding the law of oscillator absorption of radiation, our conclusion is: On the microscopic level, the probability of an oscillator absorbing radiation is proportional to the oscillator's energy distribution probability; the average energy absorbed by an oscillator is proportional to its average absorption probability. On the macroscopic level, in any frequency range, the average probability of an oscillator emitting radiation is equal to the average probability of absorbing radiation, which is the fundamental condition for thermal equilibrium radiation. This basic condition is determined by the laws of oscillator radiation emission and absorption. The laws of energy exchange between oscillators and the radiation field can be summarized as follows: The probability of an oscillator emitting radiation is proportional to the oscillator's energy level state, and the probability of an oscillator absorbing radiation is proportional to the oscillator's energy distribution probability. Although this conclusion is derived from thermal equilibrium radiation, it also applies to non-thermal equilibrium radiation.

S.4 Fundamental Laws of Thermal Radiation

Starting from the Boltzmann energy distribution law and Planck's formula, we have demonstrated the laws of oscillator radiation emission and absorption. The laws of energy exchange between oscillators and the radiation field have been well confirmed in the context of thermal equilibrium radiation. Now, based on the laws of oscillator emission and absorption of radiation, we will derive the Boltzmann energy distribution law. Historically, the Boltzmann energy distribution law developed from Maxwell's (J.C. Maxwell) gas molecular velocity distribution law, relying on the gaseous substance model. For solid materials, the Boltzmann energy distribution law has been extensively confirmed by experiments and has become one of the fundamental laws in solid-state physics, radiation physics, thermodynamics, statistical physics, and theoretical physics as a whole. However, we should also recognize that this fundamental law is an empirical law, at least in solid-state physics. The following derivation, though not a strict proof, demonstrates the inherent elegant harmony of the physical laws themselves. We begin by examining the radiation emission of oscillators.

For solid material oscillators, we know that their energy levels are uniformly spaced, and the probability of oscillator radiation emission is proportional to the oscillator's energy level state. Let the energy distribution function of material oscillators in a thermal equilibrium radiation field at temperature T be $B(\varepsilon,T)$. The distribution functions of oscillators at different energy levels are as follows:

$$B_0(\varepsilon,T), B_1(\varepsilon,T), B_2(\varepsilon,T),..., B_i(\varepsilon,T),...$$

For oscillators with frequency v, with energy levels $0hv, 1hv, 2hv,..., ihv,...$, and states 0, 1, 2,... , i, the probability of radiation emission is proportional both to the energy level state of a single oscillator and to the number of such oscillators, i.e., it is proportional to the distribution of these oscillators. $p_i(hv,T)$ represents the probability of an oscillator with frequency v and energy level i emitting radiation. Based on the radiation emission law of oscillators, we have:

$$p_i(hv,T) \propto iB_i(hv,T) \quad (4\text{-}1)$$

We multiply equation (4-1) by a coefficient C to write it as:

$$p_i(hv,T) = CiB_i(hv,T) \quad (4\text{-}2)$$

Then, the probability of radiation emission by oscillators in the frequency range $v- v+dv$ is:

$$p_i(hv,T)dv = CiB_i(hv,T)dv \tag{4-3}$$

However, we also know that the frequency v neighborhood dv interval stimulates radiation will inevitably change the energy distribution function of the oscillators in that frequency band. Since T is constant and the variable is v, it can be inferred that the probability $p_i(hv,T)dv$ is exactly the total differential of the energy distribution function $B_i(hv,T)$ under conditions of relatively stable temperature. The dynamic analysis of radiative energy is shown in Figure (4-1).

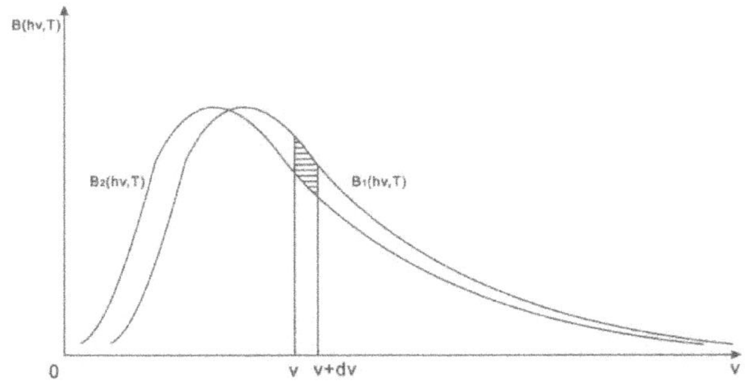

Figure (4-1) Schematic Diagram of Radiation Energy Spectrum Dynamic Analysis.

In the diagram, $B_a(hv,T)$ represents the spectrum line at time t, and $B_b(hv,T)$ represents the spectrum line at time $t+\Delta t$. Since

Δt is small, $B_b(hv,T)$ can be considered as a slight translation of $B_a(hv,T)$. As radiation causes the energy of the object to decrease, the spectrum line shifts to lower frequencies. The Δt shaded area in the diagram represents the change in the energy distribution over the time interval. Under the condition that energy is exchanged only through radiation, when $\Delta t, \Delta v$ is sufficiently small and under the condition of $\Delta T \to 0$, this change is expressed as a total differential $d(B_i(hv,T))$:

$$\begin{aligned} p_i(hv,T)dv &= d(B_i(hv,T)) \\ &= \frac{\partial B_i(hv,T)}{\partial v}dv + \frac{\partial B_i(hv,T)}{\partial T}dT_{dT\to 0} \\ &= \frac{\partial B_i(hv,T)}{\partial v}dv \end{aligned} \tag{4-4}$$

Note that the excitation of radiation by oscillators reduces energy, so we have the following relationship:

$$p_i(hv,T)dv = -dB_i(hv,T) \tag{4-5}$$

Substituting equation (4-3) into the above, we obtain:

$$\frac{dB_i(hv,T)}{B_i(hv,T)} = -Cidv \tag{4-6}$$

Integrating both sides of equation (4-6) and including the integration constant within C, we get:

$$\ln(B_i(hv,T)) = -Civ \tag{4-7}$$

Thus,

$$B_i(hv,T) = De^{-Civ} \tag{4-8}$$

where D is a normalization constant. Equation (4-8) now takes the form of the Boltzmann energy distribution function, and the coefficient C needs further analysis to determine.

The oscillator energy distribution function is the result of both oscillator radiation emission and absorption. Therefore, we must also analyze how the characteristics of oscillator absorption of radiation affect the distribution function. From the perspective of oscillator absorption of radiation, similar conclusions can be drawn. The oscillator energy distribution function is still represented by equation (4-1). Let $j_i(hv,T)$ represent the probability of an oscillator with frequency v and energy level E_{i-1} absorbing radiation. We now know that the probability of an oscillator in the energy level E_{i-1} absorbing radiation is proportional to $B_i(hv,T)$, and the probability of an oscillator in the energy level E_{i-1} absorbing radiation is exactly the probability of the oscillator absorbing i photons. Therefore, we have:

$$j_i(hv,T) \propto iB_i(hv,T) \tag{4-9}$$

For thermal equilibrium radiation, in the frequency range dv, the probabilities of radiation emission and absorption by oscillators are equal. The probability of an oscillator absorbing radiation is:

$$j_i(hv,T)dv = CiB_i(hv,T)dv \tag{4-10}$$

Oscillator absorption of radiation will also change the original distribution function, which is equivalent to reversing the relationship shown in Figure (4-1). Oscillator absorption of radiation increases energy, so we have:

$$j_i(hv,T)dv = d(B_i(hv,T)) = CiB_i(hv,T)dv \tag{4-11}$$

This is similar to equation (4-7) except for the lack of a negative sign, indicating the absorption of radiation.

In thermal equilibrium, the average probabilities of oscillator radiation emission and absorption are equal. In a non-equilibrium state, due to different temperatures, the values may vary, but the functional form of the average probability remains unchanged. It can be seen that when the temperature is fixed, the average probability is a function of frequency v. Changes in frequency in the range dv will inevitably affect the average probability. Conversely, changes in the average probability will inevitably influence the oscillator's absorption of radiation. Therefore, we should also examine how the average probability changes with the rate of change of frequency v. To do so, we differentiate equation (3-12):

$$\frac{d}{dv}\left(\frac{1}{e^{hv/kT}-1}\right) = -\frac{h}{kT}\frac{e^{hv/kT}}{\left(e^{hv/kT}-1\right)^2} \tag{4-12}$$

The derivative of the average absorption rate is negative, indicating that the absorption rate decreases with increasing frequency, which is consistent with reality. To find the relationship for the rate of change in the absorption rate, we make a room temperature approximation to equation (4-12). At room temperature, $T = 300\,K$, substituting Planck's constant and the Boltzmann constant, we get:

$$\frac{hv}{kT} = \frac{6.626\times10^{-34}v}{1.38\times10^{-23}\times3\times10^{2}} = 1.36\times10^{-13}v \qquad (4\text{-}13)$$

At room temperature, thermal energy is concentrated in the infrared region, where the characteristic frequency is large 10^{13}, $hv/kT \succ 1$, $e^{hv/kT} \succ\succ 1$. When the temperature increases, such as $T = 3000\,K$, thermal energy shifts towards the ultraviolet end, with the characteristic frequency exceeding 10^{14} , $e^{hv/kT} \succ\succ 1$. Therefore, equation (4-12) can be expressed as:

$$\frac{d}{dv}\left(\frac{1}{e^{hv/kT}-1}\right) = -\frac{h}{kT}\frac{e^{hv/kT}-1}{\left(e^{hv/kT}-1\right)^{2}}$$

$$= -\frac{h}{kT}\frac{1}{e^{hv/kT}-1} \qquad (4\text{-}14)$$

That is,

$$\frac{d\left(1/(e^{hv/kT}-1)\right)}{1/(e^{hv/kT}-1)} = -\frac{h}{kT}dv \qquad (4\text{-}15)$$

Equation (4-15) shows that the rate of change in the average absorption probability with frequency v is proportional to $\left(-h/kT\right)$. Therefore, when the frequency changes in the range dv, the probability of an oscillator absorbing radiation is also proportional to $\left(-h/kT\right)$. Equation (4-15) is equivalent to Equation (4-6), Thus, by replacing $\left(-h/kT\right)$ in equation (4-11) with C, we have:

$$\frac{d\left(B_i\left(hv,T\right)\right)}{B_i\left(hv,T\right)} = -\frac{h}{kT}dv \tag{4-16}$$

After integrating and normalizing, we obtain:

$$\ln\left(B_i\left(hv,T\right)\right) = -\frac{ihv}{kT}, \quad B_i\left(hv,T\right) = De^{-ihv/kT} \tag{4-17}$$

Equation (4-17) is the Boltzmann distribution function for oscillator energy levels.

However, we know that Planck's formula is also an experimental law. If we do not treat the oscillator's average energy, and thus the average probability, as a theoretical result of the Boltzmann energy distribution law, but instead regard it solely as an experimental law, we can say that the Boltzmann energy distribution law is the result of the energy exchange law in the radiation field. In deriving Planck's formula, the Boltzmann energy distribution function was used. Now, based on the laws of energy exchange in thermal radiation, we have derived the Boltzmann energy distribution function, which may seem like circular reasoning. However, statistical laws often have this characteristic. This is because the typical mechanical causal relationships here change, and the causal chain no longer follows a clear order, exhibiting characteristics of mutual causality. Thus, we say that the laws of oscillator energy distribution and the laws of energy exchange between oscillators and the radiation field are mutually causal conditions and two aspects of the same fundamental law. The former statically reveals the law of oscillator energy distribution, while the latter dynamically reveals the law of energy exchange between oscillators and the radiation field. Together, they are referred to as the fundamental laws of thermal equilibrium radiation. Similarly, although this conclusion is derived from thermal equilibrium radiation, the fundamental relationship holds for non-equilibrium radiation as well.

This brings us to a discussion of causality in physical processes. In classical Newtonian mechanics, causality has a clear order, where, for example, force is the cause of changes in an object's motion state, and the change in the object's motion state is the effect of the force. The causal chain in this case is clearly defined. However, in the microscopic realm, the causality between physical entities does not have the same clear order as in macroscopic objects. Heisenberg's uncertainty principle reveals that in the microscopic domain, a particle's position x and momentum p, or time t and energy E, cannot have definite values simultaneously. The fundamental reason lies in the characteristics of the causal chain order. Taking the example of the hydrogen atom's Coulomb field electron motion, the electron moves in the proton's Coulomb field, influenced by the Coulomb field, while simultaneously exciting the Coulomb field and exerting influence on the proton's Coulomb field. Therefore, the causality between the electron and the Coulomb field's state change exhibits simultaneity. This results in uncertainty in the electron's motion state. The fundamental laws of thermal radiation, as a statistical law, can be compared to the position coordinates in spacetime, while the energy exchange flow can be compared to the momentum and energy of spacetime. Their causal chain exhibits the characteristic of simultaneity.

S.5 Analysis of Thermal Equilibrium Radiation Process

We analyze the thermal equilibrium radiation process based on the fundamental laws of thermal radiation. We know that an isolated material system will eventually evolve into a thermal equilibrium system. If the energy exchange within the system occurs solely through radiation (which, in the context of the universe, is the only form of energy exchange), it is considered a thermal equilibrium radiation system. We take the example of cavity radiation, a special case. Let's assume that the cavity is adiabatic, with n types of objects inside the cavity, each with mass m_1 , m_2 ,..., m_i ,..., m_n , temperature T_1 , T_2 ,..., T_i ,..., T_n , and monochromatic emissivity $r_1(v,T_1), r_2(v,T_2),..., r_i(v,T_i),..., r_n(v,T_n)$, and the cavity has a small hole for measuring the radiation spectral density of the system inside. An adiabatic cavity is an isolated system, as shown in Figure (5-1).

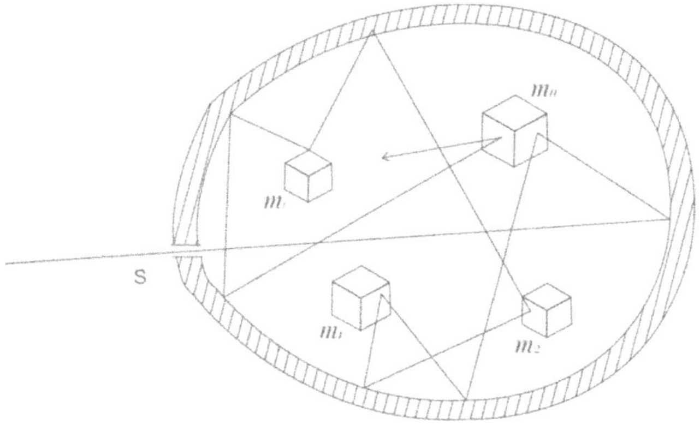

Figure (5-1) Schematic Diagram of the Thermal Equilibrium Radiation Process

For simplicity, we assume the following relationship between the temperatures of the objects inside the cavity:

$$T_1 \succ T_2 \succ \ldots \succ T_i \succ \ldots \succ T_n \tag{5-1}$$

We know that an object's ability to emit and absorb radiation depends on its structure, particularly its surface conditions (color, smoothness, etc.). Even at the same temperature, different objects have different monochromatic emissivities $r_i(v,T)$ and monochromatic irradiances $e_i(v,T)$. On the other hand, for the same object, the monochromatic emissivity $r_i(v,T)$ is closely related to the monochromatic irradiance $e_i(v,T)$. Kirchhoff's (C.R. Kirchhoff) law tells us that:

$$\frac{r_1(v,T)}{a_i(v,T)} = \frac{r_2(v,T)}{a_2(v,T)} = \dots = \frac{r_i(v,T)}{a_i(v,T)} = \frac{r_c(v,T)}{1}$$

$$= e_c(v,T) \tag{5-2}$$

In the equation, $a_i(v,T)$ represents the monochromatic absorption coefficient of an object m_i at temperature T, while $r_c(v,T), e_c(v,T)$ are the monochromatic emissivity and irradiance of an ideal black body, respectively, with $a_c(v,T)=1$ being the monochromatic absorption coefficient of the ideal black body. Equation (5-2) holds for any temperature T. This is not surprising, as it reflects the material structure and surface characteristics, which are independent of the radiation itself. Equation (5-2) can be used to express the monochromatic emissivity of different objects in terms of the monochromatic emissivity of an ideal black body. Noting that $r_c(v,T) = cu(v,T)/4$, we have:

$$r_i(v,T) = a_i r_c(v,T) = \frac{ca_i}{4} u(v,T)$$

$$= \frac{ca_i}{4} \frac{8\pi v^2}{c^3} \frac{hv}{e^{hv/kT} - 1}$$

$$= \frac{2\pi a_i v^2}{c^2} \frac{hv}{e^{hv/kT} - 1} \tag{5-3}$$

Equation (5-3) is the general formula for the monochromatic

radiance of an object. Thus, the monochromatic radiance of different objects in the cavity can be expressed in terms of the black-body monochromatic radiance. It can be anticipated that, after a sufficiently long period, the isolated system will approach thermal equilibrium. Let its equilibrium temperature be T_a. At this point, it can be inferred that the monochromatic radiance in the frequency range v near the cavity small hole dv is:

$$r(v,T_a)=\frac{c}{4}u(v,T_a)=\frac{2\pi v^2}{c^2}\frac{hv}{e^{hv/kT_a}-1} \tag{5-4}$$

The probability of an oscillator absorbing radiation is proportional to the probability of the oscillator's energy distribution. This relationship holds true for any heat exchange process. Regardless of the irradiance of the object's surface, the probability of a low-energy oscillator absorbing radiation is always greater than that of a high-energy oscillator. However, it is obvious that the amount of radiation absorbed by the object is proportional to the number of photons of that frequency irradiating the object's surface, that is, proportional to the monochromatic irradiance $e(v,T)$. The $e(v,T)$ in the cavity is determined by the average value of n of the $r_i(v,T)$ objects in the cavity. To this end, we calculate the average monochromatic radiance of the objects in the cavity. Let the number of oscillators in the object m_i with a frequency of v be N_i, and the absorption coefficient be T_i at a temperature of $a_i = a_i(v, T_i)$. Then the probability of N_i oscillators exciting radiation is equivalent to $a_i N_i$ blackbody oscillators. Therefore, the probability of object m_i exciting radiation with a frequency of v is

$$F_i(v,T_i) = a_i N_i f_i(v,T_i) = a_i N_i \frac{1}{e^{hv/kT_i} - 1} \quad (i = 1,2,\ldots,n) \qquad (5\text{-}5)$$

where $f_i(v,T_i)$ is the average probability of radiation emission for object m_i. The total probability of radiation emission for n oscillators is:

$$F(v,T_i) = \sum_{i=1}^{n} a_i N_i f_i(v,T) = \sum_{i=1}^{n} a_i N_i \frac{1}{e^{hv/kT_i} - 1} \qquad (5\text{-}6)$$

Thus, the average probability of radiation emission for these n oscillators is:

$$\overline{F}(v,T) = \frac{\sum_{i=1}^{n} a_i N_i \left(1/\left(e^{hv/kT_i} - 1\right)\right)}{\sum_{i=1}^{n} a_i N_i} \qquad (5\text{-}7)$$

For simplicity, consider a system with only two objects. Since $T_1 \succ T_2$, we have $f_1(v,T_1) \succ f_2(v,T_2)$, and we can assume

$f_2(v,T_2)=f_1(v,T_1)-\alpha$, where α is a positive real number less than $f_1(v,T_1)$. Therefore, the average probability of radiation emission for the system is:

$$\overline{F}(v,T)=\frac{a_1N_1f_1(v,T_1)+a_2N_2f_2(v,T)}{a_1N_1+a_2N_2} \tag{5-8}$$

Since:

$$\overline{F}(v,T)=\frac{a_1N_1}{a_1N_1+a_2N_2}f_1(v,T_1)+\frac{a_2N_2}{a_1N_1+a_2N_2}f_2(v,T_2)$$

$$=\frac{a_1N_1}{a_1N_1+a_2N_2}f_1(v,T_1)+\frac{a_2N_2}{a_1N_1+a_2N_2}(f_1(v,T_1)-\alpha)$$

$$=f_1(v,T_1)-\frac{a_2N_2\alpha}{a_1N_1+a_2N_2}\prec f_1(v,T_1) \tag{5-9}$$

we also have:

$$\overline{F}=\frac{a_1N_1}{a_1N_1+a_2N_2}(f_2(v,T_2)+\alpha)+\frac{a_2N_2}{a_1N_1+a_2N_2}f_2(v,T_2)$$

$$=f_2(v,T_2)+\frac{a_1N_1\alpha}{a_1N_1+a_2N_2}\succ f_2(v,T_2) \tag{5-10}$$

In summary, we have:

$$f_1(v,T_1)\succ\overline{F}(v,T)\succ f_2(v,T_2) \tag{5-11}$$

Note that the radiation probability varies continuously with temperature. In the temperature interval $[T_1,T_2]$, there must exist a temperature T_θ such that:

$$\overline{F}(v,T)=f(v,T_\theta)=\frac{1}{e^{hv/kT_\theta}-1} \quad (5\text{-}12)$$

which leads to:

$$\frac{a_1N_1}{e^{hv/kT_1}-1}+\frac{a_2N_2}{e^{hv/kT_2}-1}=\frac{a_1N_1+a_2N_2}{e^{hv/kT_\theta}-1} \quad (5\text{-}13)$$

This result can be extended to systems with multiple objects. For a multi-material system, we have:

$$f_1(v,T_1)\succ\ldots\succ f(v,T_\theta)\succ\ldots\succ f_n(v,T_n) \quad (5\text{-}14)$$

and

$$f(v,T_\theta)=\sum_{i=1}^{n}a_iN_i\frac{1}{e^{hv/kT_i}-1}/\sum_{i=1}^{n}a_iN_i$$

$$=\frac{1}{e^{hv/kT_\theta}-1} \quad (5\text{-}15)$$

Since $f(v,T_\theta)$ is the average probability of radiation emission for various objects inside the cavity, the monochromatic irradiance inside the cavity is:

$$e(v,T_\theta)=\frac{2\pi v^2}{c^2}\frac{hv}{e^{hv/kT_\theta}-1} \quad (5\text{-}16)$$

Note that the ability of objects to absorb radiation is also proportional to $a_i(v,T)$, so the probability of an object m_i absorbing radiation is:

$$j_i(v,T_i)=\frac{2\pi a_i v^2}{c^2}\frac{hv}{e^{hv/kT_\theta}-1} \quad (5\text{-}17)$$

Compared to equation (5-3), equation (5-17) provides a criterion: at any time, for objects where $T_i \succ T_\theta$, the emissivity is greater than the irradiance, meaning the object radiates more energy than it absorbs, its internal energy decreases, and its temperature tends to drop. For objects where $T_i \prec T_\theta$, the emissivity is less than the irradiance, meaning the object absorbs more energy than it radiates, its internal energy increases, and its temperature tends to rise. If we continuously measure the temperature sequence from m_1 to m_n, T_1 decreases continuously while T_n increases continuously, and we have:

$$T_{11} \succ T_{12} \succ \ldots \succ T_\theta \succ \ldots \succ T_{n2} \succ T_{n1} \qquad (5\text{-}18)$$

According to the mathematical theorem of nested intervals, the system must reach thermal equilibrium at some temperature within the interval. This equilibrium point is exactly T_θ.

At first glance, the equilibrium temperature T_θ appears to be a process quantity, changing with the progress of the equilibrium. However, a deeper analysis shows that this is not the case. If the number of objects inside the cavity is large enough and the temperature difference is sufficiently small, under these conditions, if the change in equilibrium temperature over time dt, dT, is greater than the temperature T_θ difference of the objects in the neighboring area, the temperature relationship will change as follows:

At time t

$$T_1 \succ T_2 \succ \ldots \succ T_i \succ T_\theta \succ T_{i+1} \ldots \succ T_n \tag{5-19}$$

At time $t + dt$

$$T_1 \succ T_2 \succ \ldots \succ T_\theta \succ T_i \succ T_{i+1} \succ \ldots \succ T_n \tag{5-20}$$

From the above temperature relationship, it can be seen that if T_θ is a process quantity, it is possible for object m_i to release heat at time t and absorb heat at time $t + dt$. However, the fundamental law of thermal radiation tells us that this phenomenon will not spontaneously occur. This is because any object in the thermal radiation field, if it transitions from heat emission to heat absorption, must pass through an equilibrium point where the change in heat $t + dt$ occurs, at which point the energy radiated by the object equals the energy it absorbs. Absorption of heat by the object implies that its emissivity is less than the product of irradiance and absorption coefficient. However, the irradiance inside the cavity is determined by the system and is minimally affected by changes in the emissivity of object m_i. Thus, the irradiance inside the cavity $e(v, T_\theta)$ prevents the temperature of object m_i from further decreasing, preventing the transition of m_i from emitting heat to absorbing heat.

If object m_i transitions from absorbing heat to emitting heat, it similarly passes through the equilibrium point $dQ = 0$. The transition from absorbing heat to emitting heat means that the emissivity of the object $r_i(\nu, T)$ changes from less than to greater than the irradiance $e_i(\nu, T)$. However, the irradiance $e(v, T_\theta)$ is determined by the system, and the intensity of radiation absorbed by any object in the cavity cannot exceed the irradiance $e(\nu, T_\theta)$. Therefore, for objects with an emissivity less than the irradiance, their emissivity $r_i(\nu, T)$ cannot exceed the average irradiance in the cavity $e(\nu, T_\theta)$, so the reverse phenomenon cannot occur. Similarly, it is the system's irradiance $e(\nu, T_\theta)$ that prevents the object from transitioning from absorbing heat to emitting heat.

Analysis shows that an object in the radiation field, if it is in thermal equilibrium with the radiation field, will remain in thermal equilibrium unless disturbed by external conditions. This characteristic provides significant insight into the nature of thermal equilibrium radiation. As mentioned earlier, if the number of objects inside the cavity, denoted as n, is sufficiently large and the temperature difference dT is sufficiently small, the system's equilibrium temperature T_θ will lie between T_1, T_n. Without loss of generality, assume T_θ lies between the temperature m_i, m_{i+1} of object T_i, T_{i+1}. As time progresses, the system transitions from a non-equilibrium state to a thermal equilibrium state, with the interval [T_i , T_{i+1}] narrowing. This indicates that the system's equilibrium temperature T_θ is a fixed point within the $[T_i, T_{i+1}]$ interval, and thus a fixed point of the $[T_1, T_n]$ interval. Therefore, we conclude the following about the characteristics of the thermal equilibrium process: the equilibrium temperature T_θ of the thermal equilibrium process is uniquely determined by the system's initial conditions, has the characteristics of a fixed point, and is independent of the equilibrium process.

S.6 The Fundamental Laws of Thermal Radiation and the Second Law of Thermodynamics

We have already qualitatively discussed the relationship between the fundamental laws of thermal radiation and the second law of thermodynamics. Next, we will explore the relationship between the two laws starting from the necessary and sufficient conditions for thermal equilibrium radiation in a system. In the case of an oscillator in thermal equilibrium radiation, both the average probability of excitation and absorption of radiation are given by equation (3-4). If we treat $v = v(T(t)), T = T(t)$ as a function of time t, in the thermal equilibrium radiation state, the rate of change of the average probability with time tends to zero. That is,

$$\frac{d(f(v,T))}{dt}=\frac{\partial}{\partial v}\left(\frac{1}{e^{hv/kT}-1}\right)\frac{dv}{dT}\frac{dT}{dt}+\frac{\partial}{\partial T}\left(\frac{1}{e^{hv/kT}-1}\right)\frac{dT}{dt}$$

$$=-\frac{h}{kT}\frac{e^{hv/kT}}{\left(e^{hv/kT}-1\right)^2}\frac{dv}{dT}\frac{dT}{dt}+\frac{hv}{kT^2}\frac{e^{hv/kT}}{\left(e^{hv/kT}-1\right)^2}\frac{dT}{dt}$$

$$=0 \tag{6-1}$$

Thus,

$$\frac{h}{kT}\frac{dv}{dT}=\frac{hv}{kT^2} \tag{6-2}$$

Dividing both sides by (h/kT), we obtain:

$$\frac{dv}{dT}=\frac{v}{T} \tag{6-3}$$

Thus, we have:

$$\frac{dv}{v}=\frac{dT}{T}\quad,\frac{v'dt}{v}=\frac{T'dt}{T} \tag{6-4}$$

Equation (6-4) represents the necessary and sufficient condition for thermal equilibrium radiation. The left-hand side of equation (6-4) is a function of frequency $v(t)$, while the right-hand side is a function of $T(t)$. To satisfy the equation, both sides must equal the same constant $-c$. This constant c should be related to the energy exchange between the system and the external environment. Therefore, we have two equations, which are:

$$\frac{dv}{v}=-cdt,\ \frac{dT}{T}=-cdt \tag{6-5}$$

The solution to the equation is the same negative exponential solution, which is:

$$v(t) = v_0 e^{-ct},\ T(t) = T_0 e^{-ct} \tag{6-6}$$

Equation (6-6) tells us an important conclusion: for an isolated, closed system, we must have $c = 0$, thus:

$$v(t) = v_0 e^{-ct} = v_0 \quad , T(t) = T_0 e^{-ct} = T_0 \tag{6-7}$$

Equation (6-7) indicates that for a closed system, when the thermal equilibrium radiation condition (6-4) is satisfied, the system's radiation characteristics and temperature characteristics will remain constant, maintaining a long-term thermal equilibrium radiation state. However, for an open system, $c \neq 0 \ \ (c > 0)$, when $t \to \infty$, we have:

$$v(t) = v_0 e^{-ct}_{t\to\infty} \to 0,\ \ T(t) = T_0 e^{-ct}_{t\to\infty} \to 0 \tag{6-8}$$

This indicates that any open system, as time progresses indefinitely, will eventually tend to zero radiation and evolve into an absolute zero state. This demonstrates the irreversibility of the thermal radiation process. This is precisely the core of the second law of thermodynamics.

Using the isolated system thermal equilibrium process shown in Figure (5-1) as an example, we quantitatively analyze the directional energy transfer in the thermal equilibrium radiation process. Let the temperature of the system's n objects be represented by equation (5-1), and the system's equilibrium temperature be T_θ . Thus, T_θ divides the system into high-temperature and low-temperature regions. Without loss of generality, let

$$T_1 \succ T_2 \succ \ldots \succ T_i \succ T_\theta \tag{6-9}$$

and

$$T_\theta \succ T_{i+1} \succ \ldots \succ T_{i+j} \succ \ldots T_n \tag{6-10}$$

For ease of analysis, we reduce the high-temperature region objects to a black body with a mass of M_1 and a temperature of T_1, and the low-temperature region objects to a black body with a mass of M_2 and a temperature of T_2 . The monochromatic emissivity is represented by equation (5-3)

$$r_i(v,T)=\frac{2\pi a_i v^2}{c^2}\frac{hv}{e^{hv/kT_i}-1} \qquad (i=1,2) \tag{6-11}$$

The average monochromatic irradiance inside the cavity is given by equation (5-16), and the probability of the object absorbing radiation is given by equation (5-17), that is

$$j_i(v,T)=\frac{2\pi a_i v^2}{c^2}\frac{hv}{e^{hv/kT_\theta}-1} \quad (i=1,2) \tag{6-12}$$

Because $T_1 \succ T_\theta \succ T_2$, for M_1, the emissivity is greater than the irradiance, the radiated energy is greater than the absorbed energy, the internal energy of the object decreases, and the temperature tends to decrease. For M_2, the emissivity is smaller than the irradiance, the absorbed energy is greater than the radiated energy, the internal energy of the object increases, and the temperature tends to increase.

Microscopically, the energy conservation of the radiation field inside the cavity can be expressed as:

$$\sum_{l=1}^{m}\frac{2\pi a_1 N_1 v_l^2}{c^2}\left(\frac{hv_l}{e^{hv_l/kT_1}-1}-\frac{hv_l}{e^{hv_l/kT_\theta}-1}\right)$$

$$=\sum_{l=1}^{m}\frac{2\pi a_2 N_2 v_l^2}{c^2}\left(\frac{hv_l}{e^{hv_l/kT_\theta}-1}-\frac{hv_l}{e^{hv_l/kT_2}-1}\right) \quad (l=1,2,\ldots,m) \quad (6\text{-}13)$$

In the equation, T_θ represents the system's equilibrium temperature, determined by the initial conditions. v_l denotes the thermal radiation frequency. For a non-equilibrium radiation field, we say that heat flows from high-temperature objects to low-temperature objects in an irreversible manner. This does not imply that high-temperature objects do not absorb heat, nor does it imply that low-temperature objects do not release heat; rather, it refers to the difference in the algebraic sum of the heat flowing in and out of these objects. If the outgoing heat is considered as a negative value, for M_1, the algebraic sum of the heat flowing in and out is

$$\sum_{l=1}^{m}\frac{2\pi a_1 N_1 v_l^2}{c^2}\left(\frac{hv_l}{e^{hv_l/kT_1}-1}-\frac{hv_l}{e^{hv_l/kT_\theta}-1}\right)=dQ_1 \prec 0 \qquad (6\text{-}14)$$

For M_2, the algebraic sum of the heat flowing in and out is

$$\sum_{l=1}^{m}\frac{2\pi a_2 N_2 v_l^2}{c^2}\left(\frac{hv_l}{e^{kv_l/kT_\theta}-1}-\frac{hv_l}{e^{hv_l/kT_2}-1}\right)=dQ_2 \succ 0 \qquad (6\text{-}15)$$

Under the condition of $T_1 \succ T_\theta \succ T_2$, equations (6-14) and (6-15) describe the direction of heat flow. If the thermal radiation field experiences a reversible process, at time t, $T_1 \succ T_\theta \succ T_2$, and for sufficiently small time intervals $t+dt$ (dt fully small), $T_1 \prec T_\theta \prec T_2$. The temperatures of the high and low-temperature objects will reverse. Since the emissivity $r_1(v,T)$ is determined by the energy level states of M_1 and is independent of the irradiance inside the cavity, to make $T_1 \prec T_\theta$, from equation (6-14), the necessary condition is:

$$r_1(v,T) \prec j(v,T_\theta) \qquad (6\text{-}16)$$

Similarly, to make $T_\theta \prec T_2$, from equation (6-15), the necessary condition is:

$$j(v,T_\theta) \prec r_2(v,T) \qquad (6\text{-}17)$$

However, the irradiance of the radiation field is determined by the weighted average of the emissivities of all objects inside the cavity, adhering to the condition:

$$r_1(v,T) \succ j(v,T_\theta) \succ r_2(v,T) \tag{6-18}$$

It is evident that the relationship between the emissivity and irradiance shown in equations (6-16) and (6-17) violates the fundamental condition for the generation of irradiance inside the cavity, and thus such a process cannot occur. Here, condition (6-18) restricts the flow of heat from low-temperature objects to high-temperature objects, thus determining the irreversibility of the thermal radiation process.

The irreversibility of the thermal radiation process can also be demonstrated using Kirchhoff's Law. The energy exchange in a non-equilibrium thermal radiation system follows the law of conservation of energy, where the heat released by high-temperature objects (including the high-temperature field) is equal to the heat absorbed by low-temperature objects (including the low-temperature field). While the principle of energy conservation is important, it alone does not fully describe the details of heat transfer in non-equilibrium processes. To address this, Clausius (R.E. Clausius) introduced the concept of thermal temperature ratio in his study of heat exchange processes and discovered a new state function, XXX. He proposed the concept of the entropy function, defined as

$$dS = \frac{dQ}{T} \tag{6-19}$$

and

$$\int_a^b \frac{dQ(t)}{T(t)} dt = S(b) - S(a) \tag{6-20}$$

The concept of entropy is of great significance in thermodynamic systems, and the resulting second law of thermodynamics complements the first law of thermodynamics. It reveals the irreversibility of the heat exchange process. At any moment, the heat released by a high-temperature object M_1 is dQ_1, and the heat absorbed by a low-temperature object M_2 is dQ_2. According to the principle of energy conservation, $dQ_1 = dQ_2$. However, since the temperatures of the two objects are different, their thermal temperature ratios are also different. According to this example, we have

$$\frac{dQ_1}{T_1} \prec \frac{dQ_2}{T_2} \tag{6-21}$$

We know that temperature T is a macroscopic physical quantity that describes the degree of heat of an object. It is a measure of the average internal energy of the object's microscopic components (atoms and molecules). For a general solid, the relationship between temperature and the average internal energy of atoms is

$$U^{mol} = 3N_A kT = 3RT \tag{6-22}$$

or

$$T = \frac{U^{mol}}{3R} \tag{6-23}$$

where U^{mol} is the molar internal energy density, N_A is Avogadro's constant, k is Boltzmann's constant, R is the ideal gas constant, and T is the absolute temperature. Equation (6-22) indicates that, at room temperature, the molar internal energy density of an object is directly proportional to its temperature T. Substituting equation (6-23) into equation (6-21), we get

$$\frac{dQ_1}{U_1^{mol}} \prec \frac{dQ_2}{U_2^{mpl}} \tag{6-24}$$

where U_1^{mol}, U_2^{mol} is the molar internal energy density of the object M_1, M_2, and it is known that $U_1^{mol} \succ U_2^{mol}$. Here, we transform the thermal temperature ratio into the ratio of heat to molar internal energy, referred to as the thermal-to-internal ratio. The thermal-to-internal ratio helps reveal the relationship between an object's heat exchange and the molar internal energy density between objects. Thus, equation (6-24) indicates that objects with higher internal energy densities have a lesser capacity to release heat than objects with lower internal energy densities to absorb heat. The algebraic sum of this capacity

$$\frac{dQ_2}{U_2^{mol}} - \frac{dQ_1}{U_1^{mol}} \succ 0 \tag{6-25}$$

determines the direction of heat flow. If the reverse process of the above occurs, we multiply both sides of equation (6-24) by -1, yielding

$$\frac{-dQ_2}{U_2^{mol}} \prec \frac{-dQ_1}{U_1^{mol}} \tag{6-26}$$

This equation represents that objects with lower internal energy density have a lesser capacity to release heat than objects with higher internal energy density to absorb heat. However, Kirchhoff's Law states that in any case, the ability of an object to absorb thermal radiation is directly proportional to its ability to emit thermal radiation. Therefore, if an object with a lower internal energy density, M_2, has a greater capacity to absorb heat than an object with a higher internal energy density, M_1, this is a definitive physical event, and Kirchhoff's Law limits the reversal of thermal radiation. This is because, if the ability of M_2 to absorb thermal radiation is greater than the ability of M_1 to emit thermal radiation, the ability of M_2 to emit thermal radiation at the same moment will never be less than the ability of M_1 to absorb thermal radiation. Similarly, if an object with a higher internal energy density, M_1, has a lesser ability to emit thermal radiation than an object with a lower internal energy density, M_2, then the ability of M_1 to absorb thermal radiation will never exceed the ability of M_2 to emit thermal radiation. Therefore, compared to equation (6-24), equation (6-26) violates Kirchhoff's Law and is impossible to occur.

On a microscopic level, the same principle holds. The average probability of an oscillator's excitation and absorption of radiation is expressed by the same relation. At non-equilibrium points, the average energy of the oscillators and the radiation field are in a non-equilibrium state, allowing the transfer of energy to be maintained. At equilibrium points, the average energy of the oscillators and the radiation field are in a balanced state, and any slight change in the oscillator's energy distribution will be resisted by the equilibrium state. If the oscillator develops from the equilibrium state to a state where excitation exceeds absorption, this implies that the energy distribution of the oscillator transitions to a higher energy state, disrupting the balance between the oscillator and the radiation field, which is followed by an increase in radiation energy, causing the system to return to a lower energy state. If the oscillator develops from the equilibrium state to a state where absorption exceeds excitation, this indicates that the oscillator's energy distribution shifts to a lower energy state, and the result will be an increase in absorbed energy, followed by a return to a higher energy state. Here, the fundamental laws of thermal radiation maintain the balance of the radiation field, preventing the system from evolving in the opposite direction. Therefore, we can say that the irreversibility of thermodynamic phenomena is a manifestation of the fundamental laws of thermal radiation. The second law of thermodynamics is the macroscopic manifestation of the fundamental laws of thermal radiation.

We have explored the thermal equilibrium process of isolated systems using cavity radiation as a specific form, but material systems are universally interconnected, and truly isolated systems are rare. The actual thermal equilibrium processes are more complex, but this does not hinder our understanding of the essential characteristics of thermal equilibrium processes. A closed room can be considered an isolated system, and its thermal equilibrium process can be approximately simulated by cavity radiation. However, there are always exceptions. Our universe is an excellent example of an isolated system, and it is undergoing a long-term thermal equilibrium process. However, the universe is also an open system, and its eventual outcome must evolve into a state of zero radiation and zero temperature, a super-cold and dormant state.

S.7 Description of the Thermal Equilibrium Radiation Process State

We have already explored the thermal equilibrium process of isolated systems and established that the equilibrium temperature T_a of the thermal equilibrium process is a fixed point within the temperature range. It is uniquely determined by the system's initial conditions and is independent of the equilibrium process. For an isolated system, regardless of the initial temperatures of the objects within the system, as time progresses, all objects, along with the radiation field, will approach the equilibrium temperature T_a. Next, we explore the process of this approach. For generality, let's assume there are only two types of objects in the cavity, labeled m_1, m_2, with temperatures T_1, T_2 and $T_1 \succ T_2$, respectively. We know that the system's thermal equilibrium temperature T_a lies within the temperature range $[T_1, T_2]$. It should be noted that, for the thermal equilibrium temperature T_a, we generally cannot derive it directly from the system's initial conditions. Especially for multi-object systems, it is challenging to calculate the thermal equilibrium temperature T_a based solely on initial conditions. However, T_a remains a fixed point and stays unchanged throughout the equilibrium process, which makes it easier for us to find the equilibrium temperature T_a in practical applications. For multi-object systems, we can place a thermometer on each object to record the temperature change of that object during the equilibrium process. If the temperatures of the objects are approximately continuously distributed, there will inevitably be an object whose temperature change approaches zero. At this point, the temperature of this object, T_i, will approximately approach T_a. For systems with few objects, we can place several thermometers with nearly continuous distribution in the radiation field and observe the changes in the readings during the equilibrium process. The thermometer readings that remain stable, or change slowly towards zero, indicate temperatures that approach the equilibrium temperature T_a. In summary, we assume that the equilibrium temperature T_a is a known quantity. Now, over time, T_1, T_2 will gradually approach T_a.

Regarding the approach function, we consider that for example $T_1 \to T_a$, it should satisfy the following conditions: when $t=0$, the temperature is T_1, and when $t \to \infty$, $T_1 \to T_a$. Therefore, the temperature $T_1(t)$ function of object m_1 over time should have the following form:

$$T_1(t) = T_a + (T_{10} - T_a)e^{-k_1 t} \tag{7-1}$$

where T_a is the system's thermal equilibrium temperature, and T_{10} is the measured temperature of object m_1 at the first measurement. Usually, the first measurement time is taken as the starting point for timing, i.e., at this moment, $t=0$. Clearly, equation (7-1) satisfies both the initial conditions and the final approach result. k_1 is a coefficient to be determined. Similarly, the temperature function of object m_2 is:

$$T_2(t) = T_a + (T_{20} - T_a)e^{-k_2 t} \tag{7-2}$$

It should be noted that $T_{20} \prec T_a$, and the algebraic sum inside the parentheses is less than zero. We have not set any restrictions on the exponent coefficients k_1, k_2, and whether they are constants or functions of time t still needs to be determined. We now solve for k_1 through successive approximation. From equation (7-1), we know that when $t=0$, the difference between the measured temperature and the equilibrium temperature is:

$$\Delta T_{10}=T_{10}-T_a \quad (7\text{-}3)$$

When $t=\Delta t$, if the measured temperature is T_{11}, then:

$$\Delta T_{11}=T_{11}-T_a \quad (7\text{-}4)$$

However, from equation (7-1), we know that:

$$\Delta T_{11}=(T_{10}-T_a)e^{-k_1\Delta t}=\Delta T_{10}e^{-k_1\Delta t} \quad (7\text{-}5)$$

where Δt is the unit of time, which could be seconds, hours, or any arbitrary fixed-length time period. It can be treated as unit 1. From equation (7-5), we have:

$$k_{11}=In\left(\frac{\Delta T_{10}}{\Delta T_{11}}\right) \quad (7\text{-}6)$$

Similarly, when $t=2\Delta t$, the measured temperature is T_{12}, and the temperature difference is ΔT_{12} ($\Delta T_{12}=T_{12}-T_a$), thus:

$$\Delta T_{12}=\left(T_{11}-T_a\right)e^{-k_1\Delta t}=\Delta T_{11}e^{-k_1\Delta t} \quad (7\text{-}7)$$

So,

$$k_{12}=In\left(\frac{\Delta T_{11}}{\Delta T_{12}}\right) \quad (7\text{-}8)$$

In general, when $t=i\Delta t$, the measured temperature is T_{1i}, and the temperature difference is ΔT_{1i} $(\Delta T_{1i}=T_{1i}-T_a)$, we have:

$$\Delta T_{1i} = (T_{1i-1} - T_a)e^{-k_1\Delta t} = \Delta T_{1i-1}e^{-k_1\Delta t} \tag{7-9}$$

Thus,

$$k_{1i} = In\left(\frac{\Delta T_{1i-1}}{\Delta T_{1i}}\right) \tag{7-10}$$

The series of values Δt , $2\Delta t$,..., $n\Delta t$ corresponding to the measured values over time k_1 are $k_{11}, k_{12}, ..., k_{1n}$. If these values lie on a straight line, we can represent the coefficient n with this line's equation. If these values do not lie on a straight line, a univariate linear regression method can be used to determine the functional relationship between k_1 and time t. In any case, as long as there are enough measured data, we can obtain the linear regression equation for k_1. Assuming the regression equation is:

$$k_1(t) = b_0 + b_1 t \tag{7-11}$$

Substituting into equation (7-1), the temperature evolution function for object m_1 over time is:

$$T_1(t) = T_a + (T_{10} - T_a)e^{-(b_0+b_1t)t} \tag{7-12}$$

With equation (7-12), we can calculate the temperature of object m_1 at any moment after time t_0, thereby gaining insight into the equilibrium process. Similarly, for object m_2, we can also derive the temperature function as:

$$T_2(t) = T_a + (T_{20} - T_a)e^{-(b_0+b_1t)t} \qquad (7\text{-}13)$$

Here, b_0 and b_1 are not exactly the same as the values in equation (7-12). Since $T_{20} \prec T_a$, on the graph, it will approach the equilibrium temperature T_a from below. The temperature evolution function is plotted in Figure (7-1).

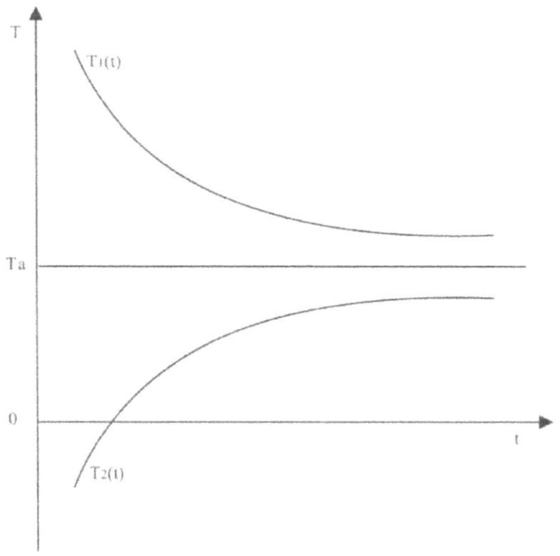

Figure (7-1) Temperature Approaching the Equilibrium Temperature During the Equilibrium Process

Typically, we need to determine the time required for the object's temperature difference to be half of its initial temperature difference. This condition is satisfied when:

$$(T_0 - T_a)e^{-(b_0+b_1t)t} = \frac{1}{2}(T_0 - T_a) \qquad (7\text{-}14)$$

Solving the equation:

$$(b_0 + b_1 t)t = In2 \tag{7-15}$$

or

$$b_1 t^2 + b_0 t - In2 = 0 \tag{7-16}$$

Solving this equation will give the time required for the temperature difference to be half of the initial temperature difference.

For multi-object systems, the situation is similar. For objects with initial temperatures greater than the equilibrium temperature T_a, their temperature will approach the equilibrium temperature from above the equilibrium temperature line. For objects with initial temperatures lower than T_a, their temperature will approach the equilibrium temperature from below the equilibrium temperature line. All temperature curves follow this negative exponential form, differing only in the initial stage, while later they tend to approach the equilibrium temperature in the same or similar form. A characteristic of this family of curves is that they do not intersect with each other. As shown in Figure (7-2).

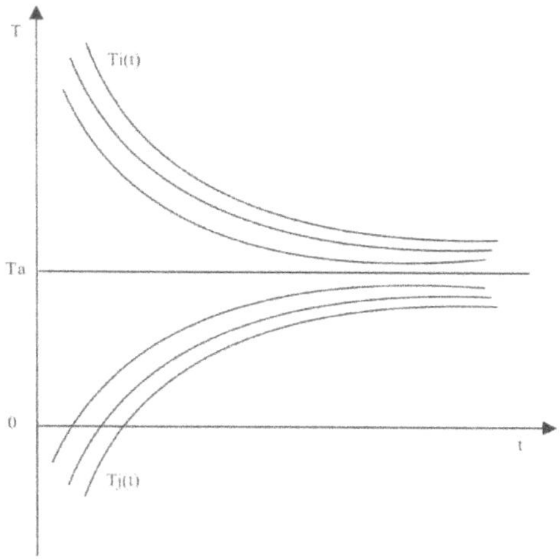

Figure (7-2) Characteristics of Object Temperature Equilibrium Curves

S.8 Conclusion

In this paper, we began with a brief review of the basic theory and concepts of thermal radiation, exploring the discovery process and significance of Planck's black-body radiation law. The focus was on the exploration and revelation of the laws governing the energy exchange between oscillators and the radiation field. Undoubtedly, this is the core topic of the book. On this foundation, we derived the Boltzmann energy distribution law, thereby providing the complete set of fundamental laws of thermal radiation. These laws encompass both energy distribution and exchange. Our conclusion is that: Material oscillators in a thermal equilibrium radiation field follow the Boltzmann energy distribution law; the probability of an oscillator emitting radiation is proportional to its energy level state; the probability of an oscillator absorbing radiation is proportional to its energy distribution probability. The Boltzmann energy distribution law and the laws governing oscillator excitation and absorption of radiation are mutually causal conditions, with the former revealing the law of oscillator energy distribution in a static sense, and the latter revealing the dynamic law of energy exchange between oscillators and the radiation field, together referred to as the fundamental laws of thermal equilibrium radiation. These conclusions, derived from thermal equilibrium radiation, also hold for non-equilibrium states.

Next, we examined the thermal equilibrium radiation process. Our conclusion regarding the thermal equilibrium process is that the equilibrium temperature of an isolated system is uniquely determined by the system's initial conditions and is independent of the equilibrium process, with the equilibrium temperature having the characteristics of a fixed point. We provided the necessary and sufficient conditions for the thermal equilibrium of independent systems. The focus was on discussing the relationship between the fundamental laws of thermal radiation and the second law of thermodynamics. In our view, the fundamental laws of thermal radiation have foundational significance. The irreversibility in heat transfer, as described by the second law of thermodynamics, is precisely the manifestation of the fundamental laws of thermal radiation. Thus, we can say that the second law of thermodynamics is the macroscopic manifestation of the fundamental laws of thermal radiation.

Our universe, from galaxies and planets to molecules and atoms, is in a state of excitation and absorption of radiation. Radiation is the most basic form of energy exchange between all things. Therefore, it can be said that the fundamental laws of thermal equilibrium radiation represent the first law of the universe. The characteristics of the fundamental laws of thermal equilibrium radiation are reflected in the concept of "regression," meaning that everything has a tendency to maintain its original state. When radiation increases, energy distribution decreases, and absorption becomes the return to the original state; conversely, the opposite occurs. In this sense, the fundamental laws of thermal equilibrium radiation inherently embody the concept of conservation.

Appendix: Derivation and Proof of the Analytical Expression for Planck's Constant

In Section S.3, when analyzing Planck's blackbody radiation law, we presented an expression for Planck's constant:

$$h = 2\pi^2 \rho_0 \varphi_0^2 c \tag{1}$$

where h represents Planck's constant, ρ_0 is the mass density of quantum space, φ_0 is the limit amplitude of quantum space, and c is the rate of radiation propagation in quantum space, i.e., the speed of light in vacuum. These basic quantities are all related to quantum space, so we must first discuss quantum space.

S.1 The Elastic Mechanical Properties of Quantum Space

Quantum space is not defined by a special meaning; in fact, it is synonymous with cosmic space. The difference is that we consider space to have a microscopic structure. That is, space is made up of elementary spaces, which are granular, have elastic deformation recovery capabilities, and are tightly bound together by mechanisms not yet fully understood, forming the macroscopic cosmic space. The cosmic space exhibits solid attributes and inherently has the characteristics of elastic deformation recovery. We call the microscopic, granular, elastic deformation-recovering cosmic space "elastic quantum space," or simply "quantum space." The basic constituent of space, which we call elementary space, is referred to as a "space particle," or "quantum particle." The way we define quantum space seems very similar to the elastic ether theory; indeed, quantum space theory is an extension and upgrade of the elastic ether theory. The difference is that we aim to quantitatively explore the physical properties of space, such as its elastic modulus k, mass density ρ, and the rest mass of the quantum particle μ_0.

1.1 Introduction to Elastic Mechanics

Indeed, some assumptions must be made. We assume that quantum space is macroscopically isotropic and a linear elastic body. In this way, we can introduce the relevant theory of linear elasticity into quantum space. According to elastic mechanics, the stress-strain relationship for an infinite, isotropic, linear elastic body is:

$$\sigma_{ij} = \lambda\varepsilon_{kk}\delta_{ij} + 2\mu\varepsilon_{ij} \tag{1-1}$$

or

$$\varepsilon_{ij} = -\frac{\lambda \sigma_{kk}}{2\mu(3\lambda + 2\mu)}\delta_{ij} + \frac{1}{2\mu}\sigma_{ij} \tag{1-2}$$

where σ_{ij} represents stress, ε_{ij} represents strain, and the indices $i\ (i = 1,2,3)$ and $j\ (j = 1,2,3)$ represent principal and directional directions, respectively. λ and μ are the Lamé constants of the isotropic elastic body. $\varepsilon_{kk} = \varepsilon_{11} + \varepsilon_{22} + \varepsilon_{33}$ and $\sigma_{kk} = \sigma_{11} + \sigma_{22} + \sigma_{33}$. δ_{ij} are the Gröbner notation symbols. For the uniaxial tensile modulus, the non-zero stress component is $\varepsilon_{kk} = \varepsilon_{11}$, and all other components are zero. Thus,

$$\varepsilon_{11} = \frac{\lambda + \mu}{\mu(3\lambda + 2\mu)}\sigma_{11}\ ,\quad \varepsilon_{22} = \varepsilon_{33} = -\frac{\lambda}{2\mu(3\lambda + 2\mu)}\sigma_{11} \tag{1-3}$$

This shows the characteristic of longitudinal stretching and circumferential shrinkage. The Young's modulus E and Poisson's ratio v are defined as:

$$E = \frac{\sigma_{11}}{\varepsilon_{11}}\ ,\quad v = -\frac{\varepsilon_{22}}{\varepsilon_{11}} = -\frac{\varepsilon_{33}}{\varepsilon_{11}} \tag{1-4}$$

The pure shear modulus G is defined as:

$$G = \frac{\sigma_{12}}{2\varepsilon_{12}}\ ,\quad \sigma_{12} = 2G\varepsilon_{12} \tag{1-5}$$

Comparing equation (1-1), we know that:

$$G = \mu \tag{1-6}$$

This is a significant characteristic of isotropic linear elastic bodies.

When a spherical object is embedded in an isotropic linear elastic body under isotropic internal pressure, the non-zero stress is:

$$\sigma_{11} = \sigma_{22} = \sigma_{33} = \frac{1}{3}\sigma_{kk} = -P \tag{1-7}$$

The bulk modulus K is defined as:

$$K = -\frac{P}{\varepsilon_{kk}} \tag{1-8}$$

From the above equations, we can derive the reciprocal relationship between the elastic moduli of the isotropic linear elastic body, which is:

$$G = \frac{E}{2(1+v)} \quad , \quad K = \frac{E}{3(1-2v)} \tag{1-9}$$

For simplicity, we set $\varepsilon_{kk} = \theta$ and note that $\mu = G$, so the constitutive equations for the isotropic linear elastic body can be expressed as:

$$\left.\begin{array}{l} \sigma_{11} = 2G\varepsilon_{11} + \lambda\theta, \sigma_{23} = \sigma_{32} = 2G\varepsilon_{23} \\ \sigma_{22} = 2G\varepsilon_{22} + \lambda\theta, \sigma_{31} = \sigma_{13} = 2G\varepsilon_{31} \\ \sigma_{33} = 2G\varepsilon_{22} + \lambda\theta, \sigma_{12} = \sigma_{21} = 2G\varepsilon_{12} \end{array}\right\} \tag{1-10}$$

and

$$\left.\begin{aligned}\varepsilon_{11} &= \frac{1}{E}[\sigma_{11} - v(\sigma_{22} + \sigma_{33})], \varepsilon_{12} = \frac{1+v}{E}\sigma_{12} \\ \varepsilon_{22} &= \frac{1}{E}[\sigma_{22} - v(\sigma_{11} + \sigma_{33})], \varepsilon_{23} = \frac{1+v}{E}\sigma_{23} \\ \varepsilon_{33} &= \frac{1}{E}[\sigma_{33} - v(\sigma_{11} + \sigma_{22})], \varepsilon_{31} = \frac{1+v}{E}\sigma_{31}\end{aligned}\right\} \quad (1\text{-}11)$$

Equations (1-10) and (1-11) are the constitutive relations for isotropic linear elastic space and serve as the foundational tools for analyzing the elastic mechanical properties of quantum space.

1.2 Characteristics of Elastic Deformation in an Infinite Isotropic Linear Elastic Body

If we embed a rigid spherical object in an infinite isotropic elastic body, the elastic space in the neighborhood of the sphere undergoes isotropic deformation, which is a typical Lamé problem. Solving such problems provides insight into analyzing the gravitational field of spherical objects in quantum space. Let us imagine a rigid spherical object embedded in an infinite isotropic linear elastic body. Using spherical coordinates, let the center of mass of the spherical object be at point o, with volume V and radius R. Under the action of internal pressure in the rigid sphere, the stress and strain in the elastic body are σ_{ij} and ε_{ij}, respectively. Based on spherical symmetry, the tangential components of the stress tensor are $\tau_{ij} = 0$, and the stress tensor is:

$$\sigma_{ij}=\begin{bmatrix}\sigma_r & \tau_{r\theta} & \tau_{r\varphi}\\ \tau_{\theta r} & \sigma_\theta & \tau_{\theta\varphi}\\ \tau_{\varphi r} & \tau_{\varphi\theta} & \sigma_\varphi\end{bmatrix}=\begin{bmatrix}\sigma_r & 0 & 0\\ 0 & \sigma_\theta & 0\\ 0 & 0 & \sigma_\varphi\end{bmatrix} i,j=(r,\theta,\varphi) \quad (1\text{-}12)$$

The strain tensor is:

$$\varepsilon_{ij}=\begin{bmatrix}\varepsilon_r & \varepsilon_{r\theta} & \varepsilon_{r\varphi}\\ \varepsilon_{\theta r} & \varepsilon_\theta & \varepsilon_{\theta\varphi}\\ \varepsilon_{\varphi r} & \varepsilon_{\varphi\theta} & \varepsilon_\varphi\end{bmatrix}=\begin{bmatrix}\varepsilon_r & 0 & 0\\ 0 & \varepsilon_\theta & 0\\ 0 & 0 & \varepsilon_\varphi\end{bmatrix} i,j=(r,\theta,\varphi) \quad (1\text{-}13)$$

The equilibrium equation for stress in spherical coordinates is:

$$\frac{\partial\sigma_r}{\partial r}+\frac{1}{r}\frac{\partial\tau_{r\theta}}{\partial\theta}+\frac{1}{r\sin\theta}\frac{\partial\tau_{r\varphi}}{\partial\varphi}+\frac{2(\sigma_r-\sigma_\theta)}{r}+\frac{\cot\theta\tau_{r\theta}}{r}=0$$

$$\frac{\partial\tau_{\theta r}}{\partial r}+\frac{1}{r}\frac{\partial\sigma_\theta}{\partial\theta}+\frac{1}{r\sin\theta}\frac{\partial\tau_{\theta\varphi}}{\partial\varphi}+\frac{3\tau_{\theta r}}{r}+\frac{\cot\theta(\sigma_\theta-\sigma_\varphi)}{r}=0 \quad (1\text{-}14)$$

$$\frac{\partial\tau_{\varphi r}}{\partial r}+\frac{1}{r}\frac{\partial\tau_{\varphi\theta}}{\partial\theta}+\frac{1}{r\sin\theta}\frac{\partial\sigma_\varphi}{\partial\varphi}+\frac{3\tau_{\varphi r}+2\tau_{\varphi\theta}\cos\theta}{r}=0$$

Note that the stress components σ_{ij} and τ_{ij} are only functions of the coordinate r, and the radial stress component $\sigma_\theta=\sigma_\varphi$. After organizing, the Lamé problem's equilibrium stress equation simplifies to:

$$\frac{\partial\sigma_r}{\partial r}+\frac{2(\sigma_r-\sigma_\theta)}{r}=0 \quad (1\text{-}15)$$

Under the action of the rigid sphere, the displacement of the elastic body is expressed by u_i $(i=r,\theta,\varphi)$. The geometric equation in spherical coordinates is:

$$\varepsilon_r = \frac{\partial u_r}{\partial r} \ , \ \varepsilon_\theta = \frac{1}{r}\frac{\partial u_\theta}{\partial \theta} + \frac{u_r}{r} \ , \ \varepsilon_\varphi = \frac{1}{r\sin\theta}\frac{\partial u_\varphi}{\partial \varphi} + \frac{u_r + \cot\theta u_\theta}{r}$$

$$\gamma_{r\theta} = \frac{1}{r}\frac{\partial u_r}{\partial \theta} + \theta\frac{\partial u_\theta}{\partial r} - \frac{u_\theta}{r} \ , \qquad \gamma_{\theta\varphi} = \frac{1}{r\sin\theta}\frac{\partial u_\theta}{\partial \varphi} + \frac{1}{r}\frac{\partial u_\varphi}{\partial \theta} - \frac{\cot\theta u_\varphi}{r}$$

$$\gamma_{\varphi r} = \frac{\partial u_\varphi}{\partial r} + \frac{1}{r\sin\theta}\frac{\partial u_r}{\partial \varphi} - \frac{u_\varphi}{r} \tag{1-16}$$

Note that $u_\theta = u_\varphi = 0$, and u_r is only a function of r, so the geometric equation simplifies to:

$$\varepsilon_r = \frac{\partial u_r}{\partial r} \ , \quad \varepsilon_\theta = \varepsilon_\varphi = \frac{u_r}{r} \tag{1-17}$$

The stress-strain relationship in spherical coordinates is:

$$\varepsilon_r = \frac{1}{E}[\sigma_r - 2v\sigma_\theta] \ , \quad \varepsilon_\theta = \frac{1}{E}[(1-v)\sigma_\theta - v\sigma_r] \tag{1-18}$$

When solving for stress, we can rewrite the equilibrium equation in the following form:

$$\sigma_\theta = \sigma_r + \frac{1}{2}r\frac{\partial \sigma_r}{\partial r} \tag{1-19}$$

Substituting equation (1-19) into equation (1-18) and simplifying, we obtain:

$$\varepsilon_r = \frac{1}{E}\left[(1-2v)\sigma_r - vr\frac{\partial \sigma_r}{\partial r}\right]$$

$$\varepsilon_\theta = \frac{1}{E}\left[(1-2v)\sigma_r + \frac{1}{2}(1-v)r\frac{\partial \sigma_r}{\partial r}\right] \tag{1-20}$$

From equation (1-17), we get:

$$\varepsilon_r = \frac{\partial(r\varepsilon_\theta)}{\partial r} \tag{1-21}$$

Substituting equation (1-20) into both sides and differentiating, we get the second-order differential equation for stress:

$$\frac{\partial^2\sigma_r}{\partial r^2} + \frac{4}{r}\frac{\partial\sigma_r}{\partial r} = 0 \tag{1-22}$$

The general solution to this equation is:

$$\sigma_r = C_1 + C_2\frac{1}{r^3} \tag{1-23}$$

where C_1 and C_2 are undetermined constants, corresponding to boundary conditions $r \to \infty$, where $\sigma_r \to 0$ and $C_1 = 0$. When $r = R$, we have $\sigma_r = -P$, so $C_2 = -PR^3$. The solution that satisfies the boundary conditions is:

$$\sigma_r = -P\frac{R^3}{r^3} \tag{1-24}$$

Substituting equation (1-24) into equation (1-19), we obtain:

$$\sigma_\theta = \sigma_\varphi = \frac{P}{2}\frac{R^3}{r^3} \tag{1-25}$$

Equations (1-24) and (1-25) indicate that under the internal pressure of the rigid sphere, the elastic restoring stress in an infinite isotropic linear elastic body is inversely proportional to r^3. Substituting equations (1-24) and (1-25) into equation (1-18), and noting that $v = -\varepsilon_\theta/\varepsilon_r$, we obtain:

$$\varepsilon_r^2 + 2\varepsilon_\theta \varepsilon_r = 0 \tag{1-26}$$

The non-zero solution of the equation is:

$$\varepsilon_r = -2\varepsilon_\theta = -2\varepsilon_\varphi \tag{1-27}$$

Poisson's ratio is:

$$v = -\frac{\varepsilon_\theta}{\varepsilon_r} = -\frac{\varepsilon_\varphi}{\varepsilon_r} = \frac{1}{2} \tag{1-28}$$

and

$$\varepsilon_{kk} = \varepsilon_r + \varepsilon_\theta + \varepsilon_\varphi = 0 \tag{1-29}$$

Equation (1-29) represents the incompressibility of the volume of an infinite isotropic elastic body. By substituting the stress solution into the stress-strain relation, we can derive the strain solution. Therefore, we have:

$$\begin{aligned} \varepsilon_r &= \frac{1}{E}\left[-P(R/r)^3 - 2v\frac{P}{2}(R/r)^3\right] \\ &= -\frac{3}{2}\frac{P}{E}(R/r)^3 \end{aligned} \tag{1-30}$$

$$\begin{aligned} \varepsilon_\theta = \varepsilon_\varphi &= \frac{1}{E}\left[(1-v)\frac{P}{2}(R/r)^3 + \frac{P}{2}(R/r)^3\right] \\ &= \frac{3}{4}\frac{P}{E}(R/r)^3 \end{aligned} \tag{1-31}$$

In fact, for an isotropic infinite elastic body, we are also unable to perform unidirectional stretching and shear tests, and thus cannot measure the Young's modulus E and shear modulus G. However, through an ideal experiment in which a rigid spherical body is embedded in an isotropic infinite linear elastic body, we can obtain the overall mechanical properties of the isotropic infinite linear elastic body. In this ideal experiment, both the expansion and shear deformations of the elastic body exhibit isotropic characteristics. Here, we introduce a new elastic modulus, E_α, defined as the isotropic Young's modulus of the infinite linear elastic body, and its relationship with the Young's modulus is given by:

$$E_\alpha = \frac{2}{3} E \tag{1-32}$$

Now, replacing E_α in equations (1-37) and (1-38) with E, we get:

$$\varepsilon_r = -\frac{P}{E_\alpha}(R/r)^3 \ , \ \varepsilon_\theta = \varepsilon_\varphi = \frac{1}{2}\frac{P}{E_\alpha}(R/r)^3 \tag{1-33}$$

Thus, the principal stress-strain relationship for the isotropic elastic body is:

$$\sigma_r = E_\alpha \varepsilon_r \ , \sigma_\theta = E_\alpha \varepsilon_\theta \ , \ \sigma_\varphi = E_\alpha \varepsilon_\varphi \tag{1-34}$$

For isotropic deformation, Poisson's ratio is $v = (1/2)$, $\varepsilon_{kk} = 0$. In this case, the stress-strain relationship for the isotropic elastic body can be written as:

$$\sigma_{ij} = \lambda \delta_{ij} \theta + 2\mu \varepsilon_{ij} = 2G\varepsilon_{ij} \tag{1-35}$$

Note that

$$2G = \frac{2}{3}E = E_{\alpha} \tag{1-36}$$

Thus, for an isotropic infinite linear elastic body under the influence of a spherical rigid body, the stress-strain relationship for the elastic body has the following important conclusion:

$$\sigma_{ij} = E_{\alpha}\varepsilon_{ij} \tag{1-37}$$

1.3 Elastic Deformation of Quantum Space

We consider that cosmic space is microscopically composed of elementary spaces—quantum particles. These quantum particles are tightly interwoven in ways that we do not yet fully understand. In quantum space, quantum particles, like atoms in a crystal lattice, have fixed positions and cannot freely move. Under the influence of physical materials and fields, quantum particles undergo deformation, vibration, and displacement from their equilibrium positions. However, within the atomic nucleus, protons, neutrons, and electrons, there are no free quantum particles. Ultimately, space has solid-state properties and is microscopically composed of quantum particles. Quantum space is illustrated in Figure (1-1). Fig. (1-1-a) shows the free quantum space, in which the neighboring holes are closely joined together. In the diagram, each hole denotes its connection with 4 joining lines. Fig. (1-1-b) is a diagram of the quantum space in the neighborhood of the atomic nucleus. The nucleus has a dislodging action on the hole as it has a definite geometric volume. The situation is similar to the dislodging action of a rigid ball on the enclosed gaseous molecules.

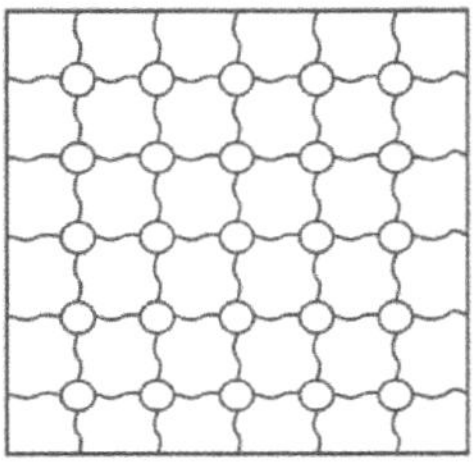

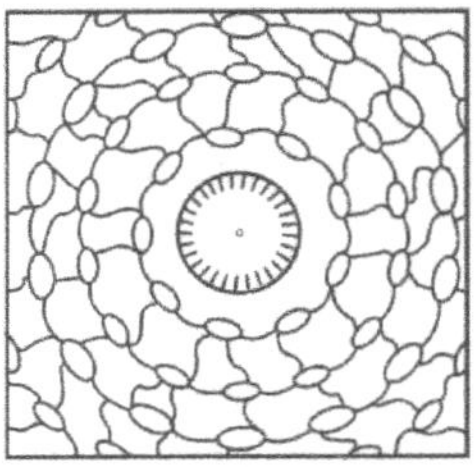

(a) (b)

Figure (1-1) Quantum Space Diagram

In the universe, physical matter, ranging from the smallest particles such as electrons, protons, neutrons, and atomic nuclei, to larger objects such as the Moon, Earth, and Sun, all exhibit spherical symmetry. Their repulsive effect on quantum space also exhibits spherical symmetry. However, because the elementary space—quantum particle—is so small, it is present everywhere in atoms and molecules, diffusing throughout them. Therefore, the objects that actually exert a repulsive effect on quantum space are atomic nuclei and external electrons, which are examples of ultra-dense geometric volumes. When discussing the repulsive effects of objects on quantum space, we always assume them to have an effective nuclear density volume. The effective nuclear density volume is represented by V_n. If the nuclear density is ρ_n, then:

$$V_n = \frac{m}{\rho_n} \tag{1-38}$$

Modern physics experiments show that the volume of an atomic nucleus is proportional to its mass. For atomic nuclei with a nucleon number A greater than 40, the nuclear radius R has a defined linear relationship with the cube root of the nucleon number A. This relationship is expressed as:

$$R = r_0 A^{1/3} \quad , \quad V_n = \frac{4}{3}\pi r^3 A \tag{1-39}$$

where r_0 ($r_0 = 1.2.--1.22$ fm) is a constant, and the atomic nucleus is approximated as a sphere. Equation (1-39) indicates that nuclear material distribution is uniform, and different atomic nuclei have the same mass density. If we take $r_0 = 1.21$ (fm), then:

$$\rho_n = \frac{m}{V_n} = \frac{1.66\times10^{-27} A}{(4\pi/3)(1.21\times10^{-15})^3 A} = 2.23\times10^{17}\ (\text{kg/m}^3)$$

We refer to objects with a nuclear density of ρ_n as super-dense materials. The constant nature of nuclear density here holds special significance. It tells us that for objects with the same mass, the effective volume is the same, and their repulsive effect on quantum space is identical. Consider quantum space embedded with a spherical object of mass m and effective volume V_n. Under the repulsive effect of this sphere, the neighboring quantum space deforms outward. To facilitate discussion, we introduce the concept of a quantum particle shell. Quantum particles in quantum space are tightly interwoven. Since quantum particles have geometric volume, we can treat quantum particles on the same plane or surface as part of the same shell. If the diameter of a quantum particle is d_0, then the distance between adjacent quantum particle shells is d_0. There is a tight connection between quantum particles and shells, and when subjected to the same repulsive force, the deformation and displacement of the particles within the same shell are the same.

Displacement of Quantum Particles Along the r Direction Now, consider a rigid spherical body with its center at o, mass m, and effective nuclear density volume V_n embedded in quantum space. Clearly, the repulsive effect of this sphere on quantum space is proportional to V_n, with kV_n and k as the proportionality coefficients. Due to this repulsive force, the neighboring quantum particle layers deform radially, and the force is transmitted layer by layer through the quantum particle shells, causing deformation in the neighboring quantum space, as shown in Figure (1-2).

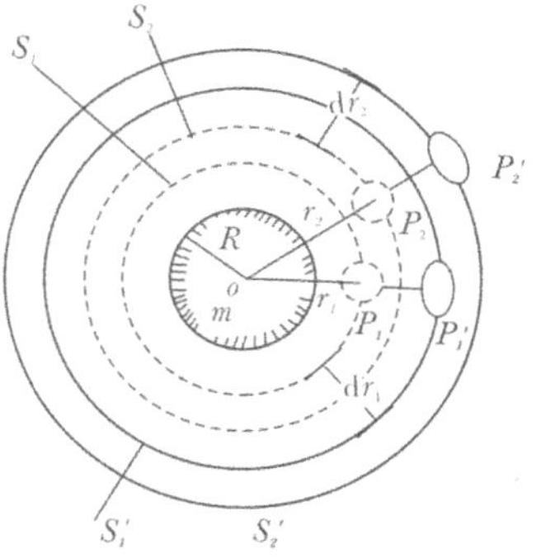

Figure (1-2) Schematic Diagram of Radial Displacement of Quantum Particle Shells in the Vicinity of a Spherical Body

The repulsive effect of the sphere is isotropic. Under this influence, the previously free-space spherical quantum particle layers S_1 and S_2 are displaced to positions S_1' and S_2', respectively. The quantum particles in the S_1' S_2' and S_2 shells correspond one-to-one. Compared to before the embedding of the sphere, the S_1' and S_2' shells now enclose a volume of V_n. Assuming that the spatial medium does not absorb or dissipate the transmitted effect, we assume that the increase in enclosed volume for any displaced quantum particle shell is V_n, compared to its original position.

The empty S_1 on the original p_1 layer is displaced to the S_1' shell p_1', with a radial displacement of $p_1p_1' = dr_1$. The empty

S_2 on the original p_2 shell is displaced to S_2' on the p_2' shell, with a radial displacement of $p_2 p_2' = dr_2$. Now the increments of the volumes enclosed by S_1' and S_2' are both V_n, so we have

$$4\pi r_1^2 dr_1 = V_n = \frac{m}{\rho_n}, \quad 4\pi r_2^2 dr_2 = V_n = \frac{m}{\rho_n}$$

In general,

$$4\pi r_i^2 dr_i = V_n = \frac{m}{\rho_n} \tag{1-40}$$

Thus, the displacement is:

$$dr = \frac{m}{4\pi \rho_n r^2} \tag{1-41}$$

Equation (1-41) means that the radial displacement dr of the empty subshell in the quantum space of the sphere neighborhood is proportional to the mass m, and inversely proportional to the square of the distance from the field point to the center of mass of the sphere.

Strain of the Quantum Particle Shell Along the r Direction

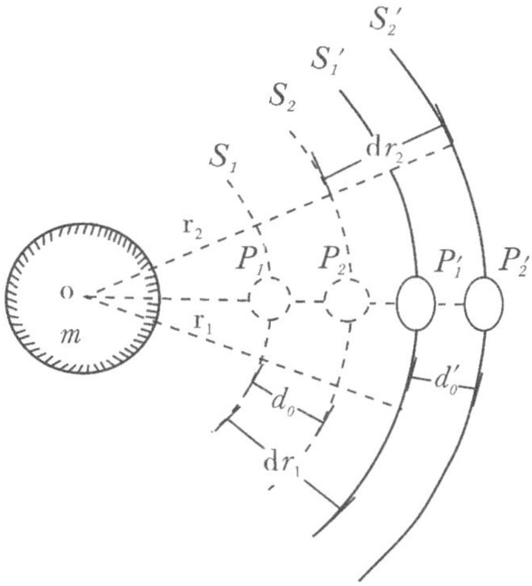

The radial deformation of the quantum particle shells in the vicinity of the sphere is illustrated in Figure (1-3).

The two adjacent spherical shell-like vacancies, denoted as . S_1, S_2, are positioned in front of the sphere, with an initial spacing of d_0. Under the influence of the sphere, the shell layers are displaced to S_1', S_2', and the vacancies p_1, p_2 are displaced to p_1', p_2', with radial displacements of dr_1 and dr_2, respectively. When the vacancy shell layers are displaced along the r direction, deformation along the r direction occurs, resulting in radial compression and circumferential expansion. The shape transforms from spherical to an ellipsoid flattened along the circumferential direction. Let the spacing of the vacancy shell layer along the r

direction after deformation be r, then d_0'. Thus, the radial change of the shell layer is given by $d_0' = p_1'p_2'$ (1-42).

$$\Delta d_0 = d_0' - d_0 = p_1'p_2' - p_1p_2$$

$$= (p_2' - p_2) - (p_1' - p_1) = dr_2 - dr_1$$

$$= \frac{m}{4\pi\rho_n(r_1 + d_0)^2} - \frac{m}{4\pi\rho_n r_1^2} = -\frac{md_0}{2\pi\rho_n r_1^3} \quad (1\text{-}42)$$

Generally,

$$\Delta d_0 = -\frac{md_0}{2\pi\rho_n r^3} \quad (1\text{-}43)$$

Equation (1-43) indicates that the radial strain of the vacancy shell layer in the vicinity of the sphere is proportional to the sphere's mass m, and inversely proportional to the cube of the distance from the field point to the center of mass. The negative sign in front of the equation indicates radial contraction.

Tangential deformation of the vacancy shell layer The radial deformation of the vacancy shell layer is accompanied by tangential deformation, as shown in Figure (1-3). Under the influence of the sphere V_n, the vacancies on the original spherical shell S_i are displaced to the S_i' shell. Regardless of how many vacancies there are on the shell S_i, they are one by one displaced to the shell S_i'.

Let the number of vacancies be N, and let σ' represent the cross-sectional area of the vacancies in the shell S_i, S_i' in the direction orthogonal to r, then we have

$$4\pi r_i^2 = N\sigma \quad , \quad 4\pi r_i'^2 = N\sigma' \tag{1-44}$$

The relative change in the cross-sectional area

$$\frac{\Delta\sigma_i}{\sigma_i} = \frac{4\pi r_i'^2 - 4\pi r_i^2}{4\pi r_i^2} = \frac{2dr}{r_i} = \frac{m}{2\pi\rho_n r_i^3} \tag{1-45}$$

If the tangential diameter of the cross-section σ_i' is denoted as d_i', generally,

$$\frac{\Delta d_0}{d_0} = \frac{2\pi r_i' - 2\pi r_i}{2\pi r_i} = \frac{dr}{r_i} = \frac{m}{4\pi\rho_n r_i^3} \tag{1-46}$$

This equation indicates that the relative change in the cross-sectional area of the vacancies is the same as the radial change, while the tangential change is half of the radial change.

At this point, regarding the quantum space under the action of a rigid sphere, we have the following conclusion about the strain tensor:

$$\varepsilon_{ij} = \begin{bmatrix} \varepsilon_r & 0 & 0 \\ 0 & \varepsilon_\theta & 0 \\ 0 & 0 & \varepsilon_\varphi \end{bmatrix} = \frac{m}{4\pi\rho_n} \begin{bmatrix} -2/r^3 & 0 & 0 \\ 0 & 1/r^3 & 0 \\ 0 & 0 & 1/r^3 \end{bmatrix} \tag{1-47}$$

That is, under the repulsive action of the sphere, the strain tensor of the quantum space is the same as the tensor form of the infinite isotropic linear elastic body Lamé problem and the form of the tidal force strain tensor in the gravitational field. According to the relationship between stress and strain, we have

$$\sigma_{ij} = 2G\varepsilon_{ii} = k\varepsilon_{ii} = \frac{km}{4\pi\rho_n}\begin{bmatrix} -2/r^3 & 0 & 0 \\ 0 & 1/r^3 & 0 \\ 0 & 0 & 1/r^3 \end{bmatrix} \tag{1-48}$$

where $k = 2G = E_\alpha$ is the isotropic Young's modulus of quantum space.

If we sum equation (1-43) from r to infinity, noting that dr contains dr/d_0 vacancy layers, we get

$$\sum_{i=1}^{\infty} \Delta d_0 = \int_r^{\infty} -\frac{md_0}{2\pi\rho_n r^3}\frac{dr}{d_0}$$

$$= \frac{m}{4\pi\rho_n r^2} = dr \tag{1-49}$$

Equation (1-49) represents the radial displacement of quantum space, which is the cumulative effect of the radial strain. Therefore, the elastic recovery effect of quantum space is the cumulative effect of radial stress. If the elastic recovery effect of quantum space is denoted by E, then at a distance of r from the center of the sphere, the elastic recovery effect per unit area of the vacancy shell layer in the orthogonal direction is

$$E = -\frac{km}{4\pi\rho_n r^2}\hat{r} = -kdr\hat{r} \tag{1-50}$$

Equation (1-50) should be the original expression of the gravitational field strength. Note that k corresponds to E_α in equation (1-37). Therefore, from Newton's gravitational formula, we obtain an important relationship, which is

$$k = E_\alpha = 4\pi\rho_n G \tag{1-51}$$

where E_α is the isotropic Young's modulus of quantum space, and G is the gravitational constant. Substituting the relevant values,

$$k = E_\alpha = 4\pi\rho_n G = 4\times3.14\times2.23\times10^{17}\times6.672\times10^{-11}$$

$$= 1.87\times10^{8}\ (\text{N/kg.m})$$

Here, k is the isotropic Young's modulus of quantum space, a characteristic quantity of quantum space. It should be noted that the gravitational field strength is an intrinsic property of the gravitational field, manifested as the cumulative effect of the stress tensor of the gravitational field, and does not require the definition of the force on a unit mass point in the gravitational field. However, we will prove that the two representations of the gravitational field strength are equivalent.

S.2 Volume Energy and Particle Transient Gravitational Field Fluctuations

2.1 Universality of Volume Energy

Since cosmic space is an elastic quantum space, and particles and ultra-dense matter exist in quantum space, causing elastic deformation of the quantum space, quantum space must inherently contain energy associated with this deformation. The energy stored in an elastic body equals the work done on it by external forces. Now, the deformation of the elastic space is triggered by the dense body V_n, so it can be considered that the elastic potential energy stored in the elastic space is proportional to V_n.

If the volume energy of the object is E_V, then $E_V = gV_n$ and g are the proportional coefficients. In the 1930s, the German physicist Wetzcoff proposed the concept of volume energy in nuclear energy research. We believe that all objects possess volume energy, and the universality of volume energy is an inevitable manifestation of elastic space. According to the principle of function, under the conditions of stable and slow loading, the strain energy density function of the unit volume of an elastic body is

$$W(\varepsilon_{ij}) = \frac{1}{2}\left(\sigma_r\varepsilon_r + \sigma_\theta\varepsilon_\theta + \sigma_\varphi\varepsilon_\varphi + \tau_{r\theta}\gamma_{r\theta} + \tau_{r\varphi}\gamma_{r\varphi} + \tau_{\theta\varphi}\gamma_{\theta\varphi}\right) \quad (2\text{-}1)$$

For quantum space, the strain energy density function $W(\varepsilon_{ij})$ is excited by the nuclear density volume V_n of the object. Given the spherical symmetry of stress and strain, we know that $\tau_{r\theta} = \tau_{r\varphi} = \tau_{\theta\varphi} = 0$, and since the loading of stress is not stable and slow, but rather an instantaneous excitation, the factor 1/2 is removed. Therefore, the elastic strain energy density excited by the sphere in quantum space is

$$W(\varepsilon_{ij}) = \sigma_r \varepsilon_r + \sigma_\theta \varepsilon_\theta + \sigma_\varphi \varepsilon_\varphi$$

$$= 2G\varepsilon_r^2 + 2G\varepsilon_\theta^2 + 2G\varepsilon_\varphi^2 \qquad (2\text{-}2)$$

Now, by substituting the tidal force stress-strain tensor of quantum space into the above equation, the volume energy becomes

$$E_V = \int_V \left(2G\varepsilon_r^2 + 2G\varepsilon_\theta^2 + 2G\varepsilon_\varphi^2\right) dV$$

$$= 2G\int_r \left(\frac{4V_n^2}{(4\pi r^3)^2} + \frac{V_n^2}{(4\pi r^3)^2} + \frac{V_n^2}{(4\pi r^3)^2} \right) 4\pi r^2 dr$$

$$= \int_R^\infty \left(\frac{12GV_n^2}{4\pi r^4} \right) dr = -\frac{4GV_n^2}{4\pi r^3}\Bigg|_R^\infty \qquad (2\text{-}3)$$

When $r \to \infty$, $1/V \to 0$, and when $r \to R$, $V \to V_n$, so

$$E_V = -\frac{(4/3)GV_n^2}{(4/3)\pi r^3}\Bigg|_R^\infty = \frac{4}{3} GV_n \qquad (2\text{-}4)$$

Equation (2-4) is interesting as it shows that volume energy is proportional to the volume of the dense body, with the proportional coefficient $g = (4/3)G$. It also indicates that quantum space has an isotropic shear modulus, denoted as G_α, i.e., $G_\alpha = (2/3)G$, so

$$E_V = \frac{4}{3}GV_n = 2G_\alpha \frac{m}{\rho_n} \qquad (2\text{-}5)$$

where

$$G_\alpha = \frac{2}{3}G = \frac{1}{3}E_\alpha = \frac{1}{3}k = 6.23\times10^7 \text{ (N/kg.m)} \qquad (2\text{-}6)$$

Equation (2-5) is the general formula for calculating volume energy. Volume energy has universality, representing the elastic potential energy of quantum space. In this sense, we have an important conclusion: the gravitational field we usually refer to is essentially a manifestation of the volume energy of an object.

From the volume energy expression, it can be seen that volume energy is a very small amount of energy. Typically, the volume energy of one kilogram of matter is

$$E_V = 2G_\alpha \frac{m}{\rho_n} = 2\times6.23\times10^7 \times\frac{1}{2.23}\times10^{-17} = 5.59\times10^{-10} \text{ (J)}$$

The volume energy of the Earth is

$$E_V = 2G_\alpha \frac{m}{\rho_n} = 2\times6.23\times10^7\times5.99\times10^{24}\times\frac{1}{2.23}\times10^{-17} = 3.35\times10^{15} \text{ (J)}$$

In conclusion, we believe that the gravitational field is an energy field, and essentially, the gravitational field is the distribution form of the object's volume energy. The energy density function is given by equation (2-2). We will later prove that the energy density given by equation (2-2) is precisely the energy component of the Einstein field equation's energy-momentum tensor.

2.2 Transient Gravitational Field Fluctuations

1) Discontinuity Characteristics of Microscopic Particle Motion The vacancy, quantum space has the ability to recover elastic deformation, which is the only physical property we assign to quantum space. The space is composed of discrete vacancies woven together, forming a solid-state lattice structure. Microscopic particles, including electrons, protons, neutrons, atomic nuclei, etc., occupy a certain geometric volume in quantum space. Thus, whenever a microscopic particle changes its spatial position, it must overcome the confinement of quantum space, exerting a repulsive effect on quantum space, while also being subjected to the elastic recovery effect of quantum space. Since the geometric size of the particle is sufficiently small, in this granular elastic space, the motion of microscopic particles exhibits a discontinuous characteristic.

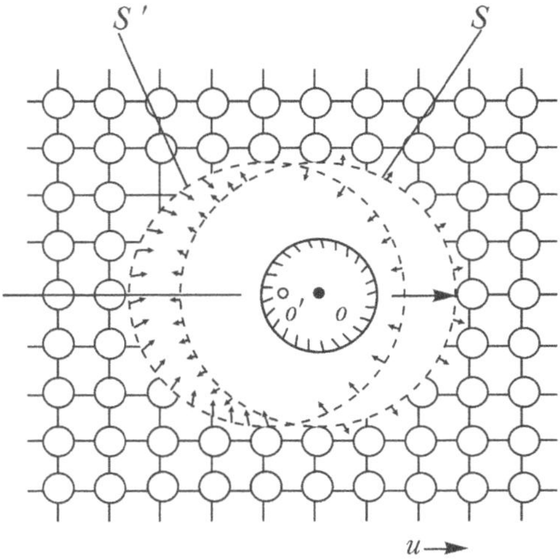

Figure (2-1) Schematic Diagram of the Discontinuity Characteristics of Microscopic Particle Motion.

As shown in Figure (2-1), let S', S represent adjacent spherical shell-like vacancy layers with a spacing of d_0. We propose that the motion of a particle from shell S' to shell S exhibits a transition characteristic. In other words, the particle has a brief "stay" in any vacancy layer, and only when the recovery effect exerted by the quantum space on the particle reaches a certain value can the particle transition from shell S' to shell S. That is, the motion of the particle exhibits discontinuity. The discontinuous, or intermittent, motion of microscopic particles is determined by the elastic granular structure characteristic of space. The discontinuity in the motion of microscopic particles forms the basis for the quantization of energy, momentum, angular momentum, and action in the microscopic domain.

2) Transient Gravitational Field Fluctuations of the Particle Assume that a particle transitions from S' to S, instantaneously forming a void centered at o', while a spherical gravitational field centered at o' recovers toward equilibrium. We know that the action is transmitted at a finite speed, and at spatial points at varying distances from the center of mass o', the recovery effect varies. Unrecovered vacancy layers will still exert force on the vacancy layers that have recovered equilibrium, causing them to deviate from their equilibrium position again, thereby inducing fluctuations in the quantum space around the region centered at o'. We refer to this fluctuation of the gravitational field, excited by the particle's volume, as transient gravitational field fluctuations. Since particles exhibit wave-particle duality, we will later demonstrate that the so-called de Broglie wave is a dynamic superposition of the transient gravitational wave of the particle.

Transient gravitational waves are spherical waves, and the propagation and vibration directions of the wave are the same. Let us take a small cylindrical element along the radial direction, with the cross-sectional area of the element $a-b$ be ds and spacing Δr. As shown in Figure (2-2).

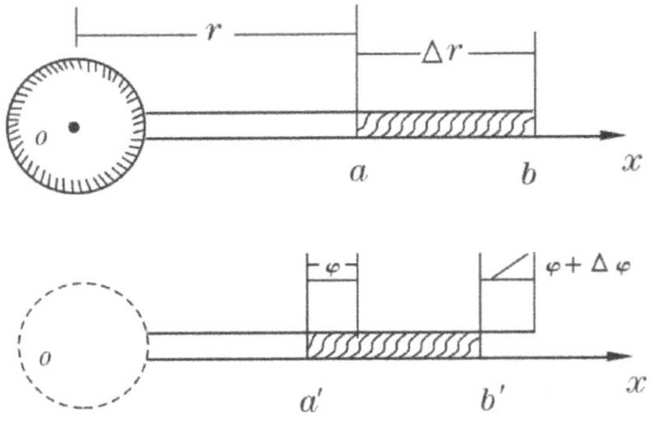

Figure (2-2) Analysis Diagram of Transient Gravitational Field Fluctuations.

When the particle at o excites a gravitational field centered at o, and the particle leaves point o, the difference in gravitational field strength at the two ends of the element a,b causes the element to vibrate. Let ΔE represent the difference in the field strength at the two ends of the element. Then,

$$\Delta E = \frac{\partial E}{\partial r}\Delta s \Delta r \tag{2-7}$$

Let the mass density of the quantum space to be determined be ρ_0, and the mass of the element is $\rho_0 \Delta s \Delta r$. The instantaneous field strength difference ΔE caused by the particle's transition causes the element to vibrate. If the vibration speed of the element is u, by applying Newton's second law to the element, we have

$$\frac{\partial E}{\partial r}\Delta s \Delta r = \frac{\partial u}{\partial t}\rho_0 \Delta s \Delta r \tag{2-8}$$

As shown in Figure (2-2), the amplitude at one end of the

element a is φ, b and at the other end it is $\varphi+\Delta\varphi$. For this longitudinal vibration of transient gravitational waves, the amplitude difference $\Delta\varphi$ is the change in the length of the element Δr. Here, $\varphi, \Delta\varphi$ is a function of r,t, and when Δt is sufficiently small,

$$\frac{\Delta\varphi}{\Delta t}=\frac{\partial\varphi}{\partial t}=u \tag{2-9}$$

Thus, equation (3-2) can be written as

$$\frac{\partial E}{\partial r}\Delta s\Delta r=\frac{\partial^2\varphi}{\partial t^2}\rho_0\Delta s\Delta r \tag{2-10}$$

In the fluctuation of the gravitational field, although the characteristics of the static field strength no longer exist, the fundamental relationship shown in equation (1-50) that the gravitational field strength is proportional to the radial relative displacement of the vacancy shell layer still holds. In the wave, the radial relative displacement of the element $a-b$ is denoted by $\Delta\varphi/\Delta r$. Thus, when $\Delta r\rightarrow dr$, we have

$$E=k\frac{\Delta\varphi}{\Delta r}=k\frac{\partial\varphi}{\partial r} \tag{2-11}$$

where k is E_α, which is the isotropic Young's modulus of quantum space. Consequently,

$$\frac{\partial E}{\partial r}=k\frac{\partial^2\varphi}{\partial r^2} \tag{2-12}$$

Substituting into equation (2-10), we obtain the equation

$$k\frac{\partial^2\varphi}{\partial r^2}\Delta s\Delta r = \frac{\partial^2\varphi}{\partial t^2}\rho_0\Delta s\Delta r \tag{2-13}$$

After simplifying, we have

$$\frac{\partial^2\varphi}{\partial r^2} = \frac{1}{(k/\rho_0)}\frac{\partial^2\varphi}{\partial t^2} \tag{2-14}$$

Equation (2-14) is the wave equation for the transient gravitational field of the particle. In this equation, (k/ρ_0) is the rate of transmission of transient gravitational waves in quantum space. On the other hand, transient gravitational waves are spherical waves, and the amplitude φ is inversely proportional to r. Noting the characteristics of the wave equation, the amplitude φ can be expressed as a linear function of gravitational potential, i.e.,

$$\varphi = lU = l\frac{km}{4\pi\rho_n r} = lEr \tag{2-15}$$

Substituting equation (2-15) into the equation for the transient gravitational field fluctuations of the particle, we have

$$\frac{\partial^2(Er)}{\partial r^2} = \frac{1}{(k/\rho_0)}\frac{\partial^2(Er)}{\partial t^2} \tag{2-16}$$

Both experimental and theoretical results indicate that the speed of gravitational waves equals the speed of light, so we can derive another characteristic quantity of quantum space, the mass density ρ_0. From $(k/\rho_0) = c^2$, we obtain

$$\rho_0 = \frac{k}{c^2} = \frac{1.87\times10^8}{9.0\times10^{16}} = 2.08\times10^{-9}\,(\mathrm{kg/m^3}) \qquad (2\text{-}17)$$

The three characteristic quantities of quantum space, k, ρ_0, c, are related through equation (2-17).

However, we know that light is a transverse wave, and its propagation rate in quantum space is c. Therefore, the elastic modulus of light is E_α, not G. This is because light is an electromagnetic wave, and photons are wave packets that are superpositions of mutually orthogonal E and H waves. Therefore, the wave and propagation of light cannot be simply understood as a transverse wave like a string. It is the combination of two orthogonal transverse waves, with its elastic modulus being $2G$. On the other hand, Equation (1-35) tells us that for quantum space, for an isotropic infinite linear elastic body, longitudinal waves and transverse waves have the same propagation speed. We have already established that $2G = E_\alpha = k$, thus explaining the identical propagation rates of electromagnetic waves and gravitational waves in terms of the elastic modulus of quantum space.

Now, by substituting the relevant values into the Planck constant formula, we can estimate the range of the extreme amplitude φ_0 of quantum space. From equation (1-1), we have

$$\varphi_0 = \sqrt{h/(2\pi^2 \rho_0 c)} = \sqrt{6.626\times10^{-34}/12.31} = 0.733\times10^{-17}(m)$$

This scale is already comparable to the fundamental length of space.

S.3 Mechanism of Mass Generation; Relationship Between Fundamental Length of Space and Nuclear Density Constant; Quantum Space Background of Newtonian Mechanics

Revealing the physical meaning of the Planck constant is undoubtedly both a meaningful and challenging proposition. However, before proving this proposition, let us first explore a more fundamental proposition—the mechanism and significance of mass generation.

3.1 Quantum Space Background of Mass Generation

The concept of mass holds immense significance in physics and all natural sciences. After several centuries of exploration, it can be said that we have yet to achieve a stunning breakthrough in core issues like the mechanism of mass generation. Newton, in his immortal work Philosophiæ Naturalis Principia Mathematica, begins with a discussion of mass. At this time, Newton had already recognized the distinction between mass and weight and discovered that density is an invariant of motion, defining the mass of an object as the product of density and volume. Newton's first law of motion, the law of inertia, states that an object remains in its state of motion unless acted upon by an external force, qualitatively explaining that inertia is the reason an object maintains its original state of motion. The second law clarifies the relationship between the net external force and the change in the object's state of motion, explaining that the change in an object's state of motion is inversely proportional to the object's inertia, and thus, its mass. Here, mass and inertia are equivalent. Apart from dynamics, Newton also discovered the law of universal gravitation, which introduces gravitational mass. Yet, mass defined under these two different concepts turns out to be remarkably equal, which is thought-provoking.

The second leap in the understanding of mass occurred when mass was recognized as an invariant of motion, marked by the establishment of the Lorentz-Einstein mass-velocity relationship and Einstein's mass-energy relation. The introduction of the mass-energy equivalence greatly changed the understanding of mass and energy.

The third phase is the ongoing exploration of the Higgs boson. Modern quantum field theory suggests that the Higgs field is the source of mass for particles, and the standard model predicts that the universe is filled with the Higgs quantum field, which stores the energy sufficient to generate all matter in the universe. Particles such as electrons and quarks acquire energy through coupling with the Higgs field, which then converts into mass according to the mass-energy relationship. Given the extraordinary achievements of the standard model and the importance of the Higgs boson within it, experiments to detect the Higgs boson have garnered significant attention. Thus, when the discovery of the Higgs boson was announced in 2013, it was met with great excitement in the theoretical community. However, there are still voices of skepticism.

We are skeptical of the Higgs mechanism. The reason is simple: since the Higgs field stores enormous energy, it would not remain so hidden that the energy density of cosmic space contains no information. A more direct reason is that we have discovered a new and better mechanism for mass generation. To explore this, we consider the following ideal experiment, as shown in Figure (3-1).

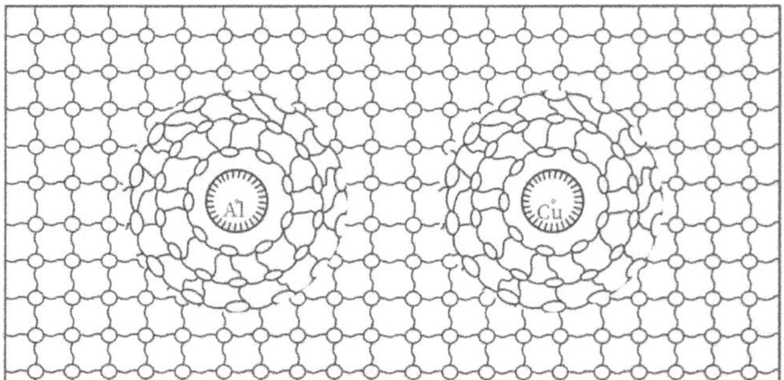

Figure (3-1) The Significance of Inertia.

Assume there is a sufficiently large isotropic linear elastic body. We embed an aluminum sphere and a copper sphere within this elastic body, requiring that the volumes of the aluminum and copper spheres be strictly equal. We know that the specific gravity of aluminum is 2.7, and that of copper is 8.9, a significant difference. However, with equal volumes, the elastic deformation induced by the elastic body will be identical for both spheres, meaning the elastic potential energy stored in the elastic body is also the same. Ignoring the effect of Earth's gravity, in order for the aluminum and copper spheres to attain the same motion state within the elastic body, the same external force is required, or in other words, they will exhibit identical inertia within the elastic body, making it impossible to distinguish between them using mechanical laws. Thus, we can say they have the same inertia and the same mass.

Cosmic space is a quantized elastic space, and the mass of any object is equal to the product of its reduced volume V_n and a constant ρ_n. The constant property of nuclear density has particular significance. If we place particles or solid matter into an elastic body for examination, it becomes easier to understand the secret of mass generation. Thus, we can conclude the following law of mass generation:

The mass of a particle is proportional to its volume, and the mass of an object is proportional to the volume V_n when compressed to nuclear density. If the object's elastic flux is kV_n or the volume energy is $2G_\alpha V_n$, the mass of the object is defined as

$$m = V_n \rho_n \tag{3-1}$$

where ρ_n is a constant related to the fundamental length of space. Basic concepts like mass cannot be derived from deeper concepts; here, mass is not derived, but defined. It is defined by the following equivalence:

$$V_n \Rightarrow kV_n \Rightarrow 2G_\alpha V_n \Rightarrow \rho_n V_n \tag{3-2}$$

Equation (3-2) contains all the secrets of the origin of mass. The constant characteristic of ρ_n plays a particularly important role here. No matter how different their chemical or physical properties, if the compressed geometric volume of two objects is the same, the geometric space they occupy in quantum space is the same. Therefore, the difficulty in changing their motion state in elastic space, and thus their inertia, is identical, meaning they have the same mass.

Quantum space has a huge effect on solid matter. However, this effect is only directly felt by super-dense objects such as atomic nuclei and electrons outside the nucleus. Atoms, molecules, and macroscopic objects formed by them cannot directly feel the effect of quantum space because their interiors are filled with vacancies. The effect of quantum space is so immense that if we could create a vacuum of 1 cubic centimeter in quantum space, it would be equivalent to moving 2.23 billion tons of matter. The mechanism of mass generation is illustrated in Figure (3-2).

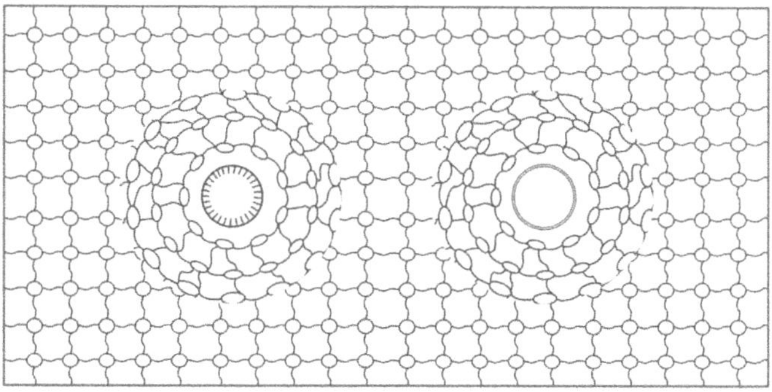

Figure (3-2) The Secret of Mass Generation.

As shown in the figure, if we somehow expel the vacancies from the geometric volume occupied by the particle, this spherical void will excite a gravitational field identical to that of the particle, thus acquiring the same mass as the particle. The true essence of mass generation lies in the volume occupied by the particle in quantum space. In modern high-energy particle collision experiments, the mass of a particle is determined by observing its energy peak. Once we better understand the quantum space background of the mass-energy relationship, we will have a new understanding of the concepts of mass and energy.

3.2 Relationship Between Fundamental Length of Space d_0 and Nuclear Density Constant ρ_n

In order to elucidate the existence of the quantum space's extreme amplitude and its possible value, we first explore the relationship between the diameter of the vacancies, also the spacing between vacancy layers ρ_n, and the nuclear density ρ_n. We hypothesize that this relationship is

$$\rho_n = \frac{1}{d_0} \tag{3-3}$$

This hypothesis is of great significance in our theory. Although this relationship is currently only a hypothesis, there are various reasons to elevate it to a theoretical status. In the schematic diagram of quantum space structure (1-1), we represent vacancies as hollow circles and the connecting lines between them represent the bonds between vacancies. Whether vacancies have an internal structure is currently unknown, and it is not a matter for discussion at this level. Therefore, what we refer to here as the diameter of a vacancy is the line thickness showing the range of its existence, which includes the vacuum field of the vacancy, i.e., the spacing between vacancy layers. If we define the fundamental length of space as l_0, we believe l_0. Just as the analysis of the secret of mass generation requires Newton's law of inertia, the relationship between the fundamental length of space d_0 and nuclear density ρ_n can be explored using Newton's equations of dynamics.

As mentioned earlier, cosmic space is quantum space, and quantum space can be depicted using vacancy layers. The motion of particles in this structured space exhibits discontinuous characteristics. Each time a particle changes its spatial position, it must expel the neighboring vacancy layers, while also being affected by the restoration effect of the balanced vacancy layers. Since the interaction between particles and quantum space is transmitted layer by layer through vacancy layers, it is a time-dependent, short-range effect with velocity. For the same vacancy shell, when expulsion and recovery occur at different times, the recovery effect always follows the expulsion effect. This is different from the interaction and reaction between the particle and quantum space. The two effects do not occur simultaneously.

The interaction between quantum space and a particle is complex. On one hand, when a particle expels the neighboring vacancy layers, it is affected by the reaction of the vacancy layers. This pair of actions is equal in magnitude but opposite in direction. The expulsion action follows the particle's instantaneous velocity. Meanwhile, the particle also experiences the elastic recovery effect of the vacancy layers. We know that the expulsion and recovery actions of the same vacancy layer are separated by a time difference of $\Delta\tau$. On the other hand, when the particle experiences the recovery effect of the vacancy layer, there is a reaction force from the vacancy layer. This pair of actions is also equal in magnitude but opposite in direction. The recovery action follows the particle's instantaneous velocity direction before $\Delta\tau$ the particle's movement. Therefore, the total effect of quantum space on the particle is determined by the vector sum of the recovery effect and the expulsion effect's reaction force (which we call the expulsion-reaction effect). The effect of the particle on quantum space is determined by the vector sum of the particle's expulsion effect and the reaction force of the recovery effect (which we call the recovery-reaction effect). Moreover, these two vector sums form a pair of actions that comply with Newton's third law. It can be anticipated that the direction of the vector sum of the effect of quantum space on the particle will not be the same as that of the vector sum of the effect of the particle on quantum space. The particle will change its motion direction. Currently, we are discussing linear motion, so there is no direction issue.

Let a particle with mass and volume m and V move in a straight line along the x axis. As shown in Figure (3-3).

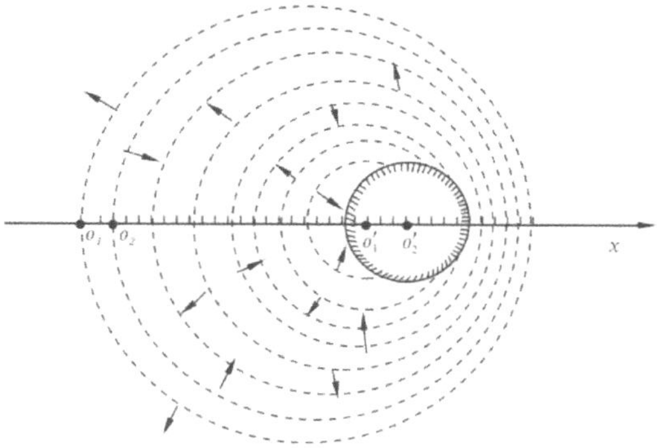

Figure (3-3) Particle Force Analysis, Schematic Diagram.

If at time $t=0$, the particle's center of mass is at o_1, and at time $t=\Delta\sigma$, the center of mass is at o_1', we define that at time o_1', the transient gravitational wave centered at o_1 tends to disappear. If at time $t=\Delta\tau$, the particle is at o_2, we define that when the particle expels the vacancy layers at o_2, the vacancy layers centered at o_1 will recover in the direction of o_1. Therefore, the particle's force situation is determined by the cooperative effect of many transient gravitational waves between $o_1 o_1'$. Let the time interval be $\Delta\sigma$, and the number of vacancy layers experienced by the particle for recovery and expulsion be n_1, n_2. Clearly, for the moving particle, n_1, n_2 is a time-dependent function. The force has an instantaneous characteristic. For a specific moment t, let dn_1/dt,

dn_2/dt represent the velocity of the recovery and expulsion-reaction forces applied by quantum space to the particle. The intensity of the force the particle experiences from quantum space is determined by the rate of change of the vector sum of the expulsion-reaction and recovery forces over time. If this vector sum is denoted by F', then

$$F' = kV\frac{d(dn_2/dt - dn_1/dt)}{dt} = kV\frac{\partial^2 n}{dt^2} \tag{3-4}$$

The essence of the force on the particle is determined by the rate of change of the expulsion-reaction and recovery vacancy layer speeds, which is a quantitative relationship. However, after rewriting the above equation, we have

$$F' = \frac{km}{\rho_n}\frac{d^2 n}{dt^2} \tag{3-5}$$

If our hypothesis holds, by substituting $d_0 (d_0 = 1/\rho_n)$, we get

$$F' = km\frac{d^2 nd_0}{dt^2} = km\frac{d^2 S}{dt^2} = kma \tag{3-6}$$

In this equation, a represents acceleration. The direction is the limit direction of the rate of change of the velocity vector sum. Here, it can be seen that the definition of acceleration reflects the inverse relationship between the fundamental length of space and the nuclear density. The originally discrete vacancy layers n are converted into the space interval S through the relationship shown in equation (3-3), thus establishing the physical meaning of acceleration. However, there is still a constant k. Equation (3-6) indicates that the force and the elastic recovery effect of quantum space are not exactly equal, although they are fundamentally the same. To clarify this point, we return to the law of universal gravitation

$$F = \frac{kmm_1}{4\pi\rho_n r^2} = m_1\left(\frac{km}{4\pi\rho_n r^2}\right) = m_1 g \tag{3-7}$$

In this equation, the acceleration g already includes k. Therefore, based on the equality of force, equation (3-6) becomes

$$F = ma \tag{3-8}$$

This is precisely the quantum space background of Newton's equations of dynamics.

Next, consider equation (3-4). When $n_2 = n_1$, $d^2n/dt^2 = 0$ the rates of change of expulsion-reaction and recovery forces are equal, and the particle experiences no force, maintaining its original motion state. If $n_2 \succ n_1$, the expulsion effect is greater than the recovery effect, and the vector sum points behind the particle. If $n_2 \prec n_1$, the recovery effect is greater than the expulsion effect, and the vector sum points along the direction of motion. It is important to emphasize that the direction of the force on the particle is always opposite to the direction of the elastic recovery effect of quantum space, or in other words, the external force on the particle is always equal and opposite to the elastic recovery effect of quantum space. In this sense, the inertial force indicated by D'Alembert's principle is precisely the elastic recovery effect of quantum space.

Interestingly, our convention regarding the fundamental length does not alter the inverse relationship between nuclear density and the fundamental length of space. For example, if 1 meter contains n vacancy layers, i.e., $nd_0 = 1(m)$, at this moment the particle's gravitational field strength is

$$E = -kdr\hat{r} = -knd_0dr\hat{r} \tag{3-9}$$

If we change the length unit to d_0, the gravitational field strength of the particle will be

$$E' = kndr\hat{r} \tag{3-10}$$

Compared to the field strength in the meter unit system, the particle's field strength has increased by a factor of n, which also means the particle's mass has increased by a factor of n, i.e., $m' = nm$. Let the nuclear density of the particle in the d_0 unit system be ρ_n', then

$$m' = V'\rho_n' \tag{3-11}$$

Thus, we have

$$m(1/d_0) = V(1/d_0)^3 \rho_n' \tag{3-12}$$

And further,

$$m = V(1/d_0)^2 \rho_n' = V(1/d_0)(\rho_n'/d_0) \tag{3-13}$$

Equation (3-13) shows that in the d_0 length unit system, the nuclear density unit is 1, but the inverse relationship between nuclear density and fundamental length d_0 still holds.

$$d_0 = \frac{1}{\rho_n} = \frac{1}{2.23\times10^{17}} = 4.48\times10^{-18}\,(\mathrm{m}) \tag{3-14}$$

To validate the relationship in equation (3-3), let us first examine the issue of the photon's rest mass. Photons propagate through quantum space, and the rest mass of a photon is the mass of the underlying space—the mass of the vacancy. With the values for space density and the diameter of the vacancies, we can calculate the rest mass of the vacancy. Let it be μ_0, then

$$\mu_0 = \frac{1}{6}\pi d_0^3 \rho_0 = \frac{1}{6}\times3.14\times\left(4.48\times10^{-18}\right)^3\times2.09\times10^{-9}$$

$$= 9.84\times10^{-62}\,(kg) \tag{3-15}$$

This magnitude is consistent with experimental results.

As we previously mentioned, the two definitions of gravitational field strength are equivalent. Now, with the concept of the origin of mass and the quantitative relationship between the fundamental length of space and nuclear density, we will continue to discuss the equivalence of the two representations of gravitational field strength. The classical expression of gravitational field strength is the force on a unit mass particle, while our expression states that gravitational field strength is the cumulative effect of stress. Stress has the characteristic of surface force. Regarding surface force, we believe the concept should be extended, as the force on an object's surface cannot be carried by an infinitely thin abstract surface. In fact, all surface forces are shell forces, carried by the object's surface lattice shell layers. Shell forces are additive, and based on this idea, according to the conclusion that gravitational field strength is the cumulative effect of the stress tensor in the vacancy shell layers, the field strength is

$$E = \frac{km}{4\pi\rho_n r^2}\hat{r} = \frac{km}{4\pi\rho_n r^2}(\sigma d_0 \rho_n)\hat{r} \qquad (3\text{-}16)$$

Let us examine the meaning of the quantities in the brackets in the above equation. Here, $\sigma = 1$ represents unit area, and d_0 represents the thickness of the shell. Therefore, the volume of the unit area shell is d_0. However, based on the secret of mass origin, when the vacuum space represented by d_0 is expelled in quantum space, it will excite a mass of size $d_0 \rho_n$, and from equation (3-3) we have $d_0\rho_n = 1$. This shows the equivalence of the two representations of gravitational field strength.

S.4 Quantum Space's Limit Amplitude; Derivation of de Broglie Relation; Analysis of Harmonic Oscillator Excited Radiation; Significance of Planck's Constant

4.1 Existence of the Limit Amplitude

The quantum space limit amplitude φ_0 calculated using Planck's constant is of the same order of magnitude as the fundamental length of space d_0 . Therefore, it is reasonable to believe that the so-called limit amplitude of quantum space is the amplitude of the vacancy at its equilibrium position. However, to understand the relationship between macroscopic and microscopic amplitudes, we consider the transverse vibration of a string. As shown in Figure (4-1).

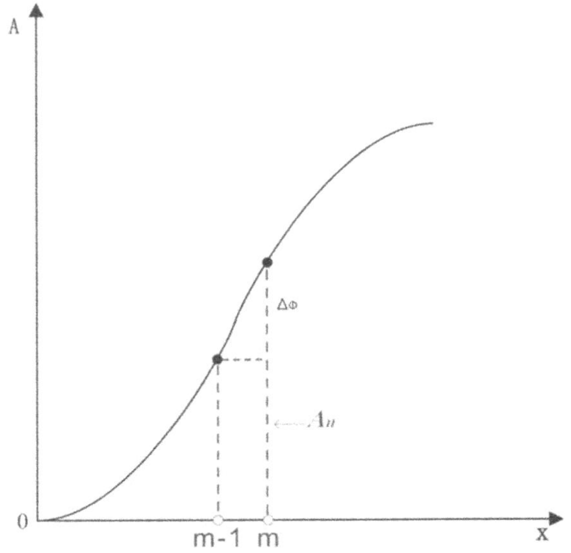

Figure (4-1) Macroscopic and Microscopic Amplitudes.

The string is composed of atoms and molecules. When the string undergoes transverse vibration, all the mass points on the string vibrate transversely. If we divide the string into numerous segments using atomic distances, we will find that each mass point vibrates with the same frequency as its adjacent mass points. For example, taking the mass point m_i in the figure, it corresponds to mass point m_{i-1} vibrating with an amplitude of $\Delta\varphi_i$. Clearly, the macroscopic amplitude of the string is the result of the accumulation of its microscopic amplitudes. If the macroscopic amplitude of the string at point x is A_x, then

$$A_x = \int_0^x \Delta\varphi(x)dx \tag{4-1}$$

For solid matter, the microscopic amplitude is represented by the lattice spacing of the material, and for quantum space, we use the amplitude of vacancies as the limit amplitude. The microscopic amplitude cannot be zero. When the amplitude of a vibration or wave approaches the limit amplitude, the vibration or wave is considered to be extinguished. That is, the condition for the extinction of vibration at the microscopic level is

$$A \to \varphi_0 \tag{4-2}$$

4.2 Kinetic Analysis of the de Broglie Wave of Microscopic Particles

Based on the discontinuity characteristics of microscopic particle motion, we analyze the origin of the de Broglie wave of particles. Generally speaking, particles not only have mass but also possess charge, and they not only excite gravitational fields but also electromagnetic fields. However, according to the principle of field action independence, gravitational fields and electromagnetic fields can be treated separately. The essence of a particle's de Broglie wave is the dynamic superposition of its transient gravitational wave. Therefore, we will only consider the particle's transient gravitational field here.

Let the particle's mass be m, its velocity be u, and it moves along the x axis in a straight line. Based on the discontinuous nature of the particle's motion, the particle has a brief "stay" in each shell layer, and only when the recovery effect reaches a certain value can it jump to the next shell layer. The particle excites a spherical gravitational field in each shell layer. This field propagates outward at the speed of light and extinguishes at the speed of light. If the radius of this spherical gravitational field is r_ε, and the propagation time of the field is Δt, then

$$\Delta t = \frac{r_\varepsilon}{c} \tag{4-3}$$

We already know that quantum space has a limit amplitude φ_0, and the amplitude of the particle's transient gravitational wave is expressed by the gravitational potential U. Therefore, r_ε, Δt satisfies the boundary condition

$$\frac{km}{4\pi\rho_n r_\varepsilon} = \frac{km}{4\pi\rho_n c\Delta t} = \varphi_0 \tag{4-4}$$

Thus, the range of the gravitational field can be represented as

$$r_\varepsilon = \frac{km}{4\pi\rho_n\varphi_0} \tag{4-5}$$

or

$$\Delta t = \frac{km}{4\pi\rho_n c\varphi_0} \tag{4-6}$$

where Δt represents the range of the transient gravitational field expressed by the wave propagation time.

We believe the particle's de Broglie wave is the dynamic superposition state of transient gravitational waves. The discontinuity feature of the particle's motion is the key to understanding the mechanism of de Broglie wave generation. As shown in Figure (2-1), if the time interval for the particle to transition from point o' to o is $\Delta\tau$, this represents the phase difference of the transient gravitational wave at point o', o as $\Delta\tau$. At the same time as the particle excites the gravitational wave at o, the gravitational field centered at point o' recovers to the center of mass. The interval between the expulsion and recovery effects at point o' is $\Delta\tau$. Thus, from a frequency perspective, the phase difference between two adjacent wave sources is exactly half of the particle's transient gravitational wave oscillation period T, i.e.,

$$\Delta\tau = \frac{1}{2}T \tag{4-7}$$

If we let the frequency of the particle's transient gravitational wave be Δv, then

$$\Delta v = \frac{1}{2\Delta\tau} \tag{4-8}$$

To understand the physical meaning of the de Broglie wave, let us consider an ideal experiment. Imagine a ball dropping onto a calm liquid surface, abbreviated as the "ball-liquid experiment." As shown in Figure (4-2).

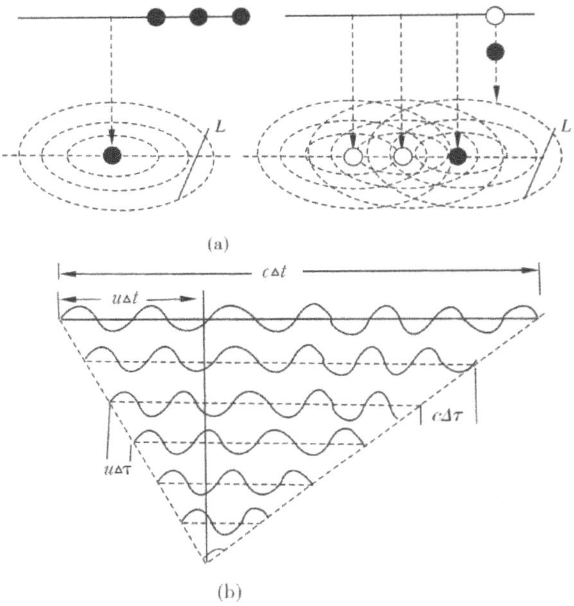

Figure (4-2) Schematic Diagram of the Ball-Liquid Experiment.

Balls are placed at equal intervals along the x axis, with a calm liquid surface beneath the x axis. Let L represent a line on the liquid surface orthogonal to the x direction. Suppose the balls on the x axis drop into the liquid surface. If only one ball drops into the liquid surface per unit of time, the liquid surface will create a wave. This wave is a circular plane wave that propagates outward at speed u. Now, if several balls drop at equal time t intervals, as shown in Figure (a), according to the independence principle of wave propagation, the wave's frequency and direction will not change due to the presence of other waves. For the ball-liquid experiment, from a frequency perspective, if one ball creates a wave with a wave number of n passing through the line L, then the number of waves created by $\Delta v'$ balls passing through the line L will be proportional to $n\Delta v'$. In other words, for an observer at point L, the frequency of the wave created by one ball is $\Delta v'$, and the frequency of the waves created by n balls is proportional to $n\Delta v'$. Considering the absorption properties of the medium, when the $n+1$the ball drops into the liquid surface, the wave created by the first ball can be neglected, and the wave passing through the line L is determined by the waves created by the n balls. Therefore, from a time interval perspective, if the time taken for the waves excited by the balls to pass to a negligible point is Δt, and the rate at which the balls drop into the liquid surface is a, let the frequency of the composite wave be v', then

$$v' \propto \Delta v' a \Delta t \tag{4-9}$$

Now we apply this conclusion to the motion of particles in quantum space. Unlike the Doppler effect of a moving wave source, the situation where a moving particle excites a de Broglie wave is similar to the composite wave in the ball-liquid experiment. Here, the particle's speed u is analogous to the number of balls dropping per unit time a, and the frequency of the transient gravitational wave Δv is similar to the frequency of a single ball's wave $\Delta v'$.

Δt represents the effective range of the wave expressed by the wave's propagation time. According to the principle of identity, the excitation and extinction time of the wave are equal, so the duration of the transient gravitational wave is $2\Delta t$. Thus, if the frequency of the particle's de Broglie wave is v, then

$$v \propto 2\Delta v u \Delta t \tag{4-10}$$

If the particle's speed is u, then $\Delta\tau = d_0/u$, and noting that $d_0 = 1/\rho_n$ we have

$$\Delta v = \frac{1}{2\Delta\tau} = \frac{1}{2}\frac{1}{d_0/u} = \frac{1}{2}\rho_n u \tag{4-11}$$

Equation (4-11) means that if the particle's speed is u, its transient gravitational wave frequency is proportional to u. This can also be understood as follows: if we interpret $\rho_n/2$ as the particle's intrinsic frequency, then its transient gravitational wave frequency is the product of the intrinsic frequency and the velocity. Substituting equation (4-11) into (4-10), we get

$$v \propto \frac{1}{2}\rho_n u^2 2\Delta t = \rho_n u^2 \Delta t \tag{4-12}$$

The frequency of the de Broglie wave is directly calculated, as shown in Figure (b). $c\Delta t$ represents the effective range of the wave, and $u\Delta t$ represents the particle's motion distance when the wave reaches the boundary. Accordingly, the number of vacancy layers for the particle is $u\Delta t/d_0$, and the contribution of the particle to the composite wave frequency at the i th shell layer is

$$\Delta v_i = \frac{1}{2}\rho_n u 2(\Delta t - i\Delta\tau) = \rho_n u(\Delta t - i\Delta\tau)\ (i = 0,1,2,\ldots,u\Delta t/d_0) \quad (4\text{-}13)$$

The contributions of each fundamental wave form an arithmetic sequence. By summing the terms, we rewrite equation (4-12) as an equation. Thus, the composite frequency of the transient gravitational wave is

$$v = \sum_{i=0}^{i=u\Delta t/d_0}(\Delta v_i)/\Delta t = \frac{1}{2}\left(\rho_n u\Delta t \times \frac{u\Delta t}{d_0}\right)/\Delta t$$

$$= \frac{1}{2}\rho_n^2 u^2 \Delta t \quad (4\text{-}14)$$

Now substituting equation (4-6) into equation (4-14), we get

$$v = \frac{1}{2}\rho_n^2 u^2 \frac{km}{4\pi\rho_n c\varphi_0} = \left(\frac{1}{2}mu^2\right)\frac{k\rho_n}{4\pi c\varphi_0}$$

$$= \frac{k\rho_n}{4\pi c\varphi_0}E_k \quad (4\text{-}15)$$

Simplifying, we have

$$E_k = \frac{4\pi c\varphi_0}{k\rho_n}v \quad (4\text{-}16)$$

Where E_k represents the particle's kinetic energy, the frequency v's prefactor quantities are all known, and substituting these into the equation, we get

$$E_k = \frac{4\times3.1416\times3\times10^8\times7.33\times10^{-18}}{1.87\times10^8\times2.23\times10^{17}}\nu$$

$$= 6.6261\times10^{-34}\nu \qquad (4\text{-}17)$$

This result is exciting: the prefactor is none other than Planck's constant h. Thus, we have

$$E_k = \frac{4\pi c\varphi_0}{k\rho_n}\nu = h\nu \quad \text{or} \quad E_k = \hbar\omega \qquad (4\text{-}18)$$

This result not only proves the existence of the quantum space limit amplitude but also provides another form of the analytical expression for Planck's constant. These two forms are independent and mutually validating.

For a free particle, $E = E_k$, under non-relativistic conditions, $E = h\nu$. The de Broglie group velocity is u, and the phase velocity is $u_\varphi = u/2$. Let E'_k represent the wave packet's kinetic energy, then $E = 2E'_k$. From $u_\varphi = \nu\lambda$ and the relationship between wave packet kinetic energy and potential energy, we obtain the expression for the free particle's momentum:

$$P = \frac{2E_k}{u} = \frac{2E'_k}{u_\varphi} = \frac{h\nu}{u_\varphi} = \frac{h}{\lambda} \quad \text{or} \quad P = \hbar k \qquad (4\text{-}19)$$

Equations (4-17) and (4-19) are the famous de Broglie relations. The analysis shows that the de Broglie wave of the particle is the dynamic superposition of the particle's transient gravitational wave, exhibiting its dynamical characteristics. For a free particle, the energy and momentum remain constant, and the frequency and wavelength of the de Broglie wave also remain constant, exhibiting the characteristics of a plane wave. However, it is fundamentally different from a plane wave propagating through a medium. First, the excitation mechanism is different: for example, electromagnetic plane waves are excited by the oscillations of a wave source, whereas de Broglie waves manifest as the dynamic superposition of transient gravitational waves. Secondly, the wave propagation speed is different: the propagation speed of electromagnetic plane waves depends on the physical properties of the electromagnetic medium, whereas the propagation speed of de Broglie waves is the particle's classical speed, and the wave packet moves with the particle at the same speed. Due to these two substantial differences, de Broglie waves cannot be represented by a plane wave function in real numbers but are represented by a complex wave function. In this way, the wave function of a free particle is:

$$\psi(r,t) = Ae^{i(k.r-\omega t)} = Ae^{-\frac{i}{\hbar}(Et-P.r)} \tag{4-20}$$

The advantage of the complex form of the wave function is that it serves as the carrier of the fundamental equation of quantum mechanics—the Schrödinger equation, from which physical quantities can be extracted using the corresponding operators. For example,

$$i\hbar\frac{\partial}{\partial t}(\psi)=E\psi\,,\quad i\hbar\nabla\psi=P\psi \tag{4-21}$$

4.3 Kinetic Analysis of Harmonic Oscillator Excited Radiation

We have analyzed the significance of the de Broglie relations based on the discontinuity features of particle motion. Now, we discuss the energy conversion of harmonic oscillator excited radiation. Microscopic particles excite de Broglie waves, and the same applies to harmonic oscillators, which oscillate around their equilibrium position. Revealing the energy relationship between the harmonic oscillator's de Broglie wave and radiation wave will help us better understand that cosmic space is a quantized elastic space.

As shown in Figure (4-3), the oscillator is at the equilibrium position o, performing oscillations with an amplitude of A_0 and a frequency of v. Unlike classical theory, we now set the harmonic oscillator in quantum space. We emphasize that microscopic particle motion has the characteristic of discontinuity, meaning that a particle has a brief stay in each vacancy shell layer, and only when the elastic recovery effect of the vacancy layer reaches a certain value can it transition to the next shell layer. For the oscillator, which is not a free particle, in addition to the discontinuity in motion, the rate of change in motion speed also has discontinuous characteristics.

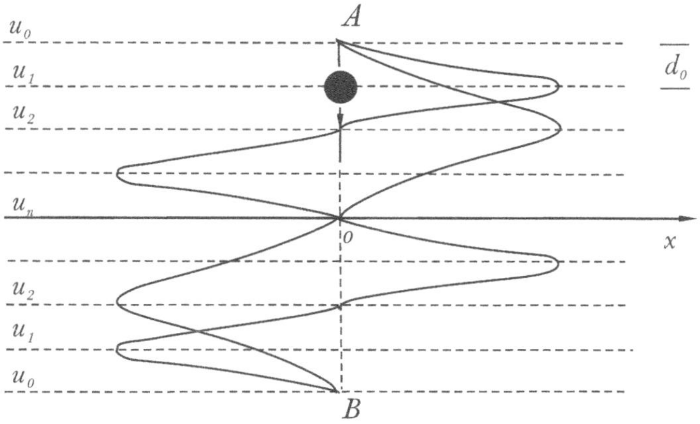

Figure (4-3) Schematic Diagram of the Harmonic Oscillator Excited Radiation.

In Figure (4-3), if the amplitude is $A_0 = nd_0$ and n represents the number of vacancy layers traversed by the oscillator from point A to the equilibrium position o, we can use symmetry. When the oscillator moves from A to B, let the velocity corresponding to the respective vacancy layers along the path of the oscillator be

$$u_0, u_1, u_2, \ldots, u_{n-1}, u_n, u_{n-1}, \ldots, u_2, u_1 \tag{4-22}$$

The kinetic energy E_k corresponding to the discontinuous speed is

$$\frac{mu_0^2}{2}, \frac{mu_1^2}{2}, \frac{mu_2^2}{2}, \ldots, \frac{mu_{n-1}^2}{2}, \frac{mu_n^2}{2}, \frac{mu_{n-1}^2}{2}, \ldots, \frac{mu_2^2}{2}, \frac{mu_1^2}{2} \tag{4-23}$$

According to the de Broglie relation ($E = hv$), the corresponding de Broglie wave frequencies are

$$\frac{mu_0^2}{2h}, \frac{mu_1^2}{2h}, \frac{mu_2^2}{2h}, \ldots, \frac{mu_{n-1}^2}{2h}, \frac{mu_n^2}{2h}, \frac{mu_{n-1}^2}{2h}, \ldots, \frac{mu_2^2}{2h}, \frac{mu_1^2}{2h} \quad (4\text{-}24)$$

If the time for the oscillator to have a discontinuous velocity is

$$d\tau_0, d\tau_1, d\tau_2, \ldots, d\tau_{n-1}, d\tau_n, d\tau_{n-1}, \ldots, d\tau_2, d\tau_1 \quad (4\text{-}25)$$

Temporarily neglecting the $u_0 = 0$ state, the number of oscillation cycles related to the oscillator's kinetic energy as it moves from A to B is

$$N = \frac{mu_1^2}{2h} d\tau_1 + \frac{mu_2^2}{2h} d\tau_2 + \ldots + \frac{mu_n^2}{2h} d\tau_n + \ldots$$

$$+ \frac{mu_2^2}{2h} d\tau_2 + \frac{mu_1^2}{2h} d\tau_1 \quad (4\text{-}26)$$

Notice that

$$d\tau_1 + d\tau_2 + \ldots + d\tau_n + \ldots + d\tau_2 + d\tau_1 = \frac{1}{2} T = \frac{1}{2v} \quad (4\text{-}27)$$

In this equation, T, v represents the oscillator’s oscillation period and frequency. Thus, the contribution of kinetic energy to the de Broglie wave frequency is

$$f_k = \frac{2N}{T} = \frac{1}{hT} \left(mu_1^2 d\tau_1 + mu_2^2 d\tau_2 + \ldots + mu_n^2 d\tau_n + \ldots \right.$$

$$\left. + mud_2^2 \tau_2 + mu_1^2 d\tau_1 \right) \quad (4\text{-}28)$$

On the other hand, from the setting of the discontinuity in the oscillator's motion speed, the time interval for the velocity of u_i degrees to traverse the vacancy layers is

$$d\tau_i = d_0/u_i \tag{4-29}$$

Substituting this into the previous equation, using the de Broglie momentum-wavelength relation $mu = h/\lambda$, we get

$$f_k = \frac{d_0}{hT}(mu_1 + mu_2 + ... + mu_n + ... + mu_2 + mu_1)$$

$$= \frac{d_0}{hT}\left(\frac{h}{\lambda_1} + \frac{h}{\lambda_2} + ... + \frac{h}{\lambda_n} + ... + \frac{h}{\lambda_2} + \frac{h}{\lambda_1}\right)$$

$$= d_0 v\left(\frac{1}{\lambda_1} + \frac{1}{\lambda_2} + ... + \frac{1}{\lambda_n} + ... + \frac{1}{\lambda_2} + \frac{1}{\lambda_1}\right) \tag{4-30}$$

Similar to a particle in a one-dimensional potential well, based on the principle of wave packet generation, the de Broglie wavelength of the harmonic oscillator, denoted as $\lambda_1, \lambda_2, ..., \lambda_i, ..., \lambda_n$, cannot be arbitrarily chosen. For any wavelength λ_i, it must satisfy

$$i\frac{\lambda_i}{2} = A_0 \qquad (i = 1,2,...,n) \tag{4-31}$$

Substituting equation (4-31) into equation (4-30), we get

$$f_k = \frac{d_0 v}{2}\left(\frac{1}{A_0} + \frac{2}{A_0} + ... + \frac{n}{A_0} + ... + \frac{2}{A_0} + \frac{1}{A_0}\right)$$

$$= \frac{d_0 v}{2A_0}(1 + 2 + ... + n + ... + 2 + 1)$$

$$=\frac{d_0 v}{2A_0}n^2=\frac{nd_0}{2A_0}nv \tag{4-32}$$

Noticing that $nd_0 = A_0$, the contribution of the oscillator's kinetic energy to the de Broglie wave frequency is

$$f_k=\frac{1}{2}nv \tag{4-33}$$

In this equation, v is the frequency of the harmonic oscillator. Thus, the harmonic oscillator's kinetic energy is

$$E_k=hf_k=\frac{1}{2}nhv \tag{4-34}$$

Where n is the quantum number, which represents the number of vacancy layers the oscillator has traversed.

Unlike a free particle, a harmonic oscillator has potential energy. In the above analysis, we only statically utilized the discrete velocities and kinetic energy of the harmonic oscillator at different points. We should also consider the contribution of potential energy to the harmonic oscillator's de Broglie wave frequency. The contribution of potential energy to the particle's de Broglie wave frequency depends on the change in potential energy. The harmonic oscillator system has zero-point kinetic energy, and the change in potential energy can be expressed by the kinetic energy at that point. Therefore, the contribution of potential energy to the harmonic oscillator's de Broglie wave frequency is

$$f_p = \frac{1}{2} n v \tag{4-35}$$

The potential energy of the harmonic oscillator is

$$E_p = h f_p = \frac{1}{2} n h v \tag{4-36}$$

Thus, the total energy of the harmonic oscillator is

$$E = E_k + E_p = n h v \tag{4-37}$$

However, the above equation does not include the state of $u_0 = 0$, which means the zero-point energy E_0 of the oscillator is not included. Zero-point energy can be calculated by determining the probability, and in the limiting case, if the amplitude is $A_0 = d_0$, that is, $n = 1$, then $E_1 = hv$. Since the oscillator is not at rest, $u_0 = 0$ represents one of the oscillator's states. Due to the discontinuity in velocity, at this moment, the oscillator chooses a velocity of u_0, with probabilities of one-half for each value of u_1. According to the Boltzmann energy distribution, the number of oscillators in the E_1 state is far fewer than those in the ground state. That is, between E_0, E_1, the probability of the oscillator choosing E_0 approaches 1, and therefore

$$E_0 = (1)\left(\frac{1}{2}\right) h v = \frac{1}{2} h v \tag{4-38}$$

Equation (4-41) shows that the zero-point energy of the harmonic oscillator is a probabilistic value. By combining equations (4-37) and (4-38), we have

$$E = \left(n + \frac{1}{2} \right) hv \tag{4-39}$$

This result is fully consistent with quantum mechanics but is derived directly through analysis. It demonstrates the dynamic characteristics of quantum mechanics. We know that the harmonic oscillator emits electromagnetic radiation, and the energy of the oscillator as expressed by equation (4-39) excites n photons with a frequency of v. The mechanism of electromagnetic radiation is shown in Figure (4-4). Equation (4-39) reveals that the radiation energy of the harmonic oscillator is a transformed form of the de Broglie wave energy.

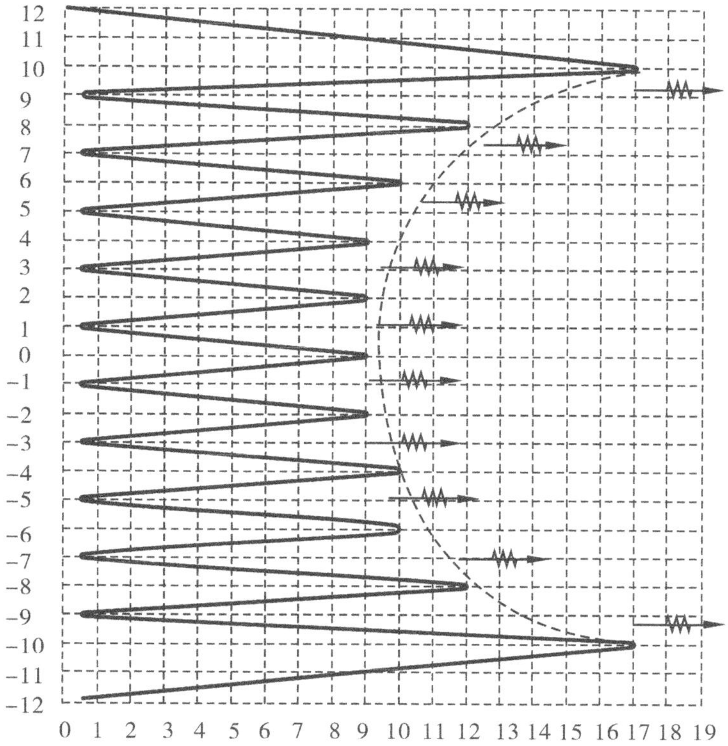

Figure (4-4) Schematic Diagram of Harmonic Oscillator Excited Photon.

We have already explored the generation mechanism of the de Broglie wave for a free particle and analyzed the energy conversion process of linear harmonic oscillator excited electromagnetic radiation. Now, we qualitatively discuss the fundamental issues of quantum mechanics, Schrödinger's equation, and wave functions. The time-independent Schrödinger equation is

$$i\hbar\frac{\partial\psi(r,t)}{\partial t}=-\frac{\hbar^2}{2\mu}\nabla^2\psi(r,t)+U(r)\psi(r,t) \tag{4-40}$$

Where

$$\nabla^2=\frac{\partial^2}{\partial x^2}+\frac{\partial^2}{\partial y^2}+\frac{\partial^2}{\partial z^2} \tag{4-41}$$

For the time-independent wave function, the wave function $\psi(r,t)$ can be decomposed as

$$\psi(r,t)=\varphi(r)f(t) \tag{4-42}$$

Substituting equation (4-42) into equation (4-40), we obtain

$$i\hbar\frac{1}{f(t)}\frac{\partial f(t)}{\partial t}=\frac{1}{\varphi(r)}\left(-\frac{\hbar^2}{2\mu}\nabla^2+U(r)\right)\varphi(r) \tag{4-43}$$

Since the left side of the equation is a function of t and the right side is a function of r, for both sides to remain equal, they must equal the same constant. On the other hand, the left side is the energy operator, and the right side is the Hamiltonian operator. Therefore, this constant should be the system's energy E, i.e.,

$$i\hbar\frac{\partial f(t)}{\partial t}=Ef(t) \tag{4-44}$$

And

$$-\frac{\hbar^2}{2\mu}\nabla^2\varphi(x)+U(x)\varphi(x)=E\varphi(x) \tag{4-45}$$

Energy has many forms. For a single particle, the universal forms of energy include rest energy, kinetic energy, potential energy, volume energy, and electrostatic energy for charged particles, denoted by E_0, E_k, E_p, E_V and E_e, respectively. It is worth discussing what specific energy forms correspond to the energy $i\hbar(\partial/\partial t)$ of the energy operator E. In terms of magnitude, $E_0 = m_0 c^2$ is much greater than the other four energy forms. However, through the analysis of the free particle de Broglie wave generation mechanism and the exploration of the energy conversion of the linear harmonic oscillator's radiation excitation, it is clear that the energy associated with the excitation of the de Broglie wave only depends on the particle's motion velocity and is independent of other energy forms. Therefore, the selection condition for energy forms is

$$\frac{\partial E}{\partial u} = \frac{\partial}{\partial u}\left(E_0 + E_k + E_p + E_V + E_e\right)$$

$$= \frac{\partial}{\partial u}\left(\mu c^2 + \frac{1}{2}\mu u^2 + U(x) + 2G_a\frac{\mu}{\rho_n} + \frac{1}{4\pi\varepsilon_0}\frac{e^2}{2r_e}\right)$$

$$\neq 0 \tag{4-46}$$

Where μ is the particle's rest mass, u is the velocity, G_a is the isotropic shear modulus of space, ρ_n is the nuclear density, ε_0 is the vacuum dielectric constant, and e and r_e are the charge and classical radius of the particle, respectively. The first and fourth terms in the parentheses are constants of motion, and the fifth term, the charge e , is an invariant of motion. Thus, the equation can be written as

$$\frac{\partial E}{\partial u}=\frac{\partial}{\partial u}\left(\frac{1}{2}\mu u^2+U(x)\right)=\mu u+\frac{\partial U(x)}{\partial x}\frac{\partial x}{\partial u}$$

$$=p-F\frac{udt}{\partial u}=p-F\frac{1}{(\ln u)'}\frac{dt}{du}\neq 0 \qquad (4\text{-}47)$$

Equation (4-47) means that, in a potential field, when the particle's velocity u approaches a constant, and thus $\ln u$ approaches a constant, the force $F=0$, the second term in the equation describing the particle's momentum change, is zero. At this point, the particle corresponds to the zero-point potential energy state, approaching a free particle. When the velocity u is not constant, the second term describes the increase in the particle's momentum, which instantaneously changes the particle's momentum and kinetic energy, thereby altering the frequency and wavelength of the particle's de Broglie wave. Therefore, under non-relativistic conditions, the quantity corresponding to the operator $i\hbar(\partial/\partial t)$ can only be the Hamiltonian, i.e., $E=H$. For instance for particles in a gravitational field,assuming an velocity $u_0=0$,Then equation (4-47) evolves into:

$$\frac{\partial E}{\partial u}=p-F\frac{1}{(\ln u)'a}=p-F\frac{1}{(\ln(gdt))'g}$$

$$=p-Fdt=p-dp \tag{4-48}$$

The above formula is also applicable to charged particles in the Coulomb field.

To explain the meaning of the wave function, we use an example. We consider the motion of a particle in a one-dimensional infinite potential well. As shown in Figure (4-5).

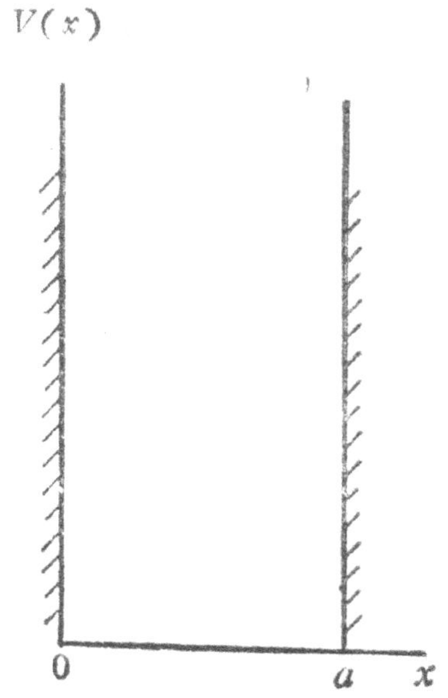

Figure (4-5) Infinite Deep Square Potential Well

The function form of the infinite deep potential well is

$$U(x)=\begin{cases}0 & 0 \prec x \prec a \\ \infty & x \prec 0, x \succ a\end{cases} \tag{4-49}$$

Outside the potential well, $U=\infty$, the time-independent Schrödinger equation is

$$\frac{1}{\varphi(x)}\frac{\partial^2\varphi(x)}{\partial x^2}=\frac{2\mu}{\hbar^2}\big(U(x)-E\big)\to\infty \tag{4-50}$$

Equation (4-50) must satisfy the conditions of continuity, single-valuedness, and boundedness for the wave function $\varphi(x)$, whose second derivative is bounded. If the above condition holds, we must have

$$\varphi(x)\equiv 0 \tag{4-51}$$

This represents that the wave function outside the potential well is always zero, and the particle cannot cross the potential boundary.

Inside the potential well ($0 \prec x \prec a$), $U(x)=0$, the time-independent Schrödinger equation is

$$\frac{\partial^2\varphi(x)}{\partial x^2}=-\frac{2\mu E}{\hbar^2}\varphi(x) \tag{4-52}$$

Let

$$k^2=\frac{2\mu E}{\hbar^2} \tag{4-53}$$

Then the equation simplifies to

$$\frac{\partial^2 \varphi(x)}{\partial x^2} = -k^2 \varphi(x) \tag{4-54}$$

The solution to equation (4-54) is

$$\varphi(x) = A\sin(kx + \delta) \tag{4-55}$$

A, δ is an undetermined constant. Using the boundary condition for the continuity of the wave function, we have

$$\varphi(0) = 0 \tag{4-56}$$

$$\varphi(a) = 0 \tag{4-57}$$

From equation (4-56), we get $\delta = 0$, and from equation (4-57), we get

$$\sin ka = 0 \tag{4-58}$$

Thus, we have

$$ka = n\pi \qquad (n = 1,2,3,\ldots) \tag{4-59}$$

Substituting into equation (4-53), we get

$$E = E_n = \frac{\hbar^2 \pi^2 n^2}{2\mu a^2} \qquad (n = 1,2,3,\ldots) \tag{4-60}$$

Equation (4-60) represents that the energy of the particle in the infinite deep potential well is discrete, with a discontinuous spectrum.

The normalized wave function corresponding to the energy level E_n is

$$\varphi_n(x) = A_n \sin\left(\frac{n\pi}{a}x\right) = \sqrt{\frac{2}{a}}\sin\left(\frac{n\pi}{a}x\right) \tag{4-61}$$

Thus, the time-dependent form of the wave function is

$$\psi_n(x,t) = \sqrt{\frac{2}{a}}\sin\left(\frac{n\pi}{a}x\right)e^{-\frac{i}{\hbar}E_n t}$$

$$= \sqrt{\frac{2}{a}}\frac{\left(e^{in\pi x/a} - e^{-in\pi x/a}\right)}{2i}e^{-\frac{i}{\hbar}E_n t} \tag{4-62}$$

On the other hand, we know that the de Broglie wave is also no exception. The condition for the wave to form a standing wave inside the potential well is

$$n\frac{\lambda_n}{2} = a \qquad (n = 1,2,3,\ldots) \tag{4-63}$$

As shown in Figure (4-6).

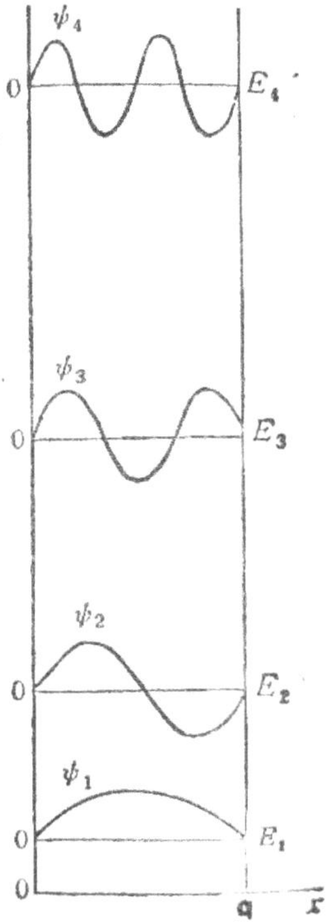

Figure (4-6) Standing Wave in an Infinite Deep Potential Well.

Substituting equation (4-58) into equation (4-57), we get

$$\psi_n(x,t) = -\frac{i}{\sqrt{2a}}\left(e^{i2\pi x/\lambda_n} - e^{-i2\pi x/\lambda_n}\right)e^{(-i/\hbar)E_n t}$$

$$= -\frac{i}{\sqrt{2a}}\left(e^{(-i/\hbar)(E_n t-(h/\lambda_n)x)} - e^{(-i/\hbar)(E_n t-(h/\lambda_n)x)}\right)$$

$$= -\frac{i}{\sqrt{2a}}\left(e^{(-i/\hbar)(E_n t-p_n.x)} - e^{(-i/\hbar)(E_n t+p_n.x)}\right) \qquad (4\text{-}64)$$

Equation (4-59) is of significant meaning, clearly indicating that the wave function inside the potential well, corresponding to energy E_n, is a superposition of two de Broglie waves with energy ψ_n and momentum E_n, moving in opposite directions. However, inside the potential well, there is only one particle, so equation (4-59) represents that the wave function p_n is the superposition of the primary wave (the wave moving with the particle) and the secondary wave (the wave that continues after the particle leaves). This characteristic is common to all constrained state particle wave functions.

4.4 Physical Meaning of Planck's Constant

The above analysis shows that the de Broglie wave of a particle has a fundamental meaning, and in equation (4-39), the quantum number of the harmonic oscillator represents the number of photons with excitation energy hv . The underlying significance is the number of vacancy layers traversed by the oscillator. Figure (4-6) shows a schematic diagram of the quantization of the oscillator’s amplitude, indicating that the amplitude in the microscopic domain is also quantized.

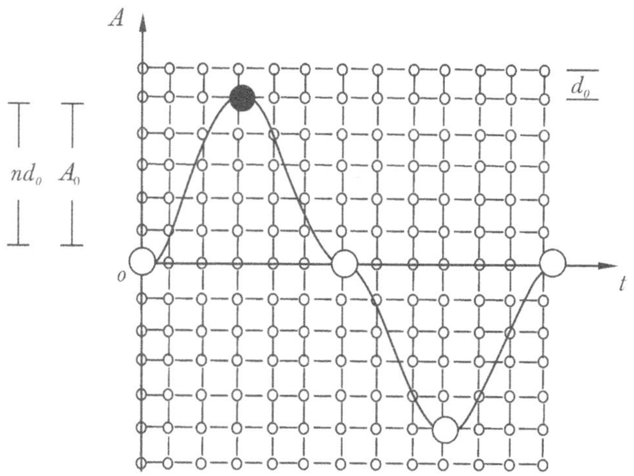

Figure (4-6) Schematic Diagram of Oscillator Amplitude Quantization

Classical theory gives the oscillation energy of the harmonic oscillator

$$E = \frac{1}{2} m A_0^2 \omega^2 \tag{4-65}$$

In this equation, m, A_0, ω represents the mass, amplitude, and angular frequency of the oscillator. The oscillation of the oscillator causes vibrations in the surrounding quantum space, and both have the same frequency. According to classical wave energy density and classical electromagnetic field theory, we can express the energy density of radiation for quantum space at unit wavelength as

$$\varepsilon = \frac{1}{2} \rho_0 \varphi_0^2 \omega^2 \lambda \tag{4-66}$$

Where ρ_0, φ_0 is the mass density and limit amplitude of quantum space, and ω is the angular frequency. During one period, the vibration in quantum space propagates a distance of one wavelength λ. If the oscillator vibrates with amplitude A_0 and excites $2A_0$ radiation with frequency n in the interval ω, It can be expressed by the average energy of n wave packets with wavelength λ and frequency ω in the radiation field, i.e.,

$$E = \frac{1}{2} m A_0^2 \omega^2 = n\left(\frac{1}{2} \rho_0 \varphi_0^2 \omega^2 \lambda\right)$$

$$= n(2\pi^2 \rho_0 \varphi_0^2 v^2 \lambda) = n(2\pi^2 \rho_0 \varphi_0^2 c)v$$

$$= nhv \qquad (4\text{-}67)$$

The numbers in the brackets represent Planck's constant. Its physical meaning is the energy density of a single electromagnetic radiation, or the energy density of one oscillation of a wave packet within a wavelength range. With this, we have completed the proof of the analytical expression for Planck’s constant.

S.5 Mass-Energy Relation and Dark Energy; Gravitational Field Energy-Momentum Tensor; Einstein Field Equation Variational Solution; Dynamics of Cosmic Expansion

5.1 Mass-Energy Relation

After proving the analytical expression for Planck's constant, it seems appropriate to conclude the discussion. However, given the in-depth analysis of quantum space, discussing the mass-energy relation, volume energy, and related issues in general relativity will enhance our understanding of the properties of space.

One of the core relationships in physics, the mass-energy relation, is

$$E = mc^2 \tag{5-1}$$

Under non-relativistic conditions, m represents the rest mass. We used the discussion of the behavior of the aluminum and copper balls in the elastic body to demonstrate that inertia is a property of elastic space, thus revealing the secret of the origin of mass. Regarding energy, we directly define it as a physical quantity that characterizes an object's ability to disturb quantum space. A moving particle excites a de Broglie wave, a charged particle excites an electromagnetic field, a harmonic oscillator excites electromagnetic radiation, a gravitational field is the manifestation of volume energy, and thermal energy is a measure of the average kinetic energy of atoms and molecules—all related to the deformation, disturbance, vibration, and fluctuation of quantum space. The difference in the energy of objects reflects their ability to disturb quantum space. Considering the mass-energy relation in quantum space, we have

$$E = mc^2 = m\frac{E_\alpha}{\rho_0} = E_\alpha \frac{m}{\rho_0} \tag{5-2}$$

Where E_α is the isotropic Young's modulus of quantum space. Equation (5-2) represents that the annihilation of an object's mass releases the underlying space—vacancies, generating a space with a volume of (m/ρ_0) in quantum space. This produces a quantity of

$$E_p = 2G_\alpha \frac{m}{\rho_0}$$

Volume energy, but since $E_\alpha = 3G_\alpha$, equation (5-2) can be expressed in two parts, i.e.,

$$E = E_\alpha \frac{m}{\rho_0} = \frac{2E_\alpha}{3}\frac{m}{\rho_0} + \frac{E_\alpha}{3}\frac{m}{\rho_0}$$

$$= 2G_\alpha \frac{m}{\rho_0} + G_\alpha \frac{m}{\rho_0} \tag{5-3}$$

Equation (5-3) indicates that 66.67% of the energy from mass annihilation is converted into volume energy, which drives the expansion of the cosmic space, while 33.33% is converted into radiation energy, remaining in the cosmic space, exciting isotropic fluctuations in quantum space. Equation (5-4) represents the final form of mass-energy conversion.

For particles with non-negligible velocity, the energy relation of mass annihilation is.

$$E = mc^2 = m_0 c^2 \left(1 - u^2/c^2\right)^{-1/2}$$

$$= m_0 c^2 \left(1 + \frac{1}{2}\frac{u^2}{c^2} + \ldots\right) = m_0 c^2 + \frac{1}{2} m_0 u^2 + \ldots$$

$$= 2G_\alpha \frac{m_0}{\rho_0} + G_\alpha \frac{m_0}{\rho_0} + h\nu \tag{5-4}$$

The first term in the above equation represents the volume energy produced by the annihilation of the particle, the second term represents the isotropic radiation energy excited by the particle's mass annihilation, and the third term represents the energy of the particle's de Broglie wave.

From a cosmological perspective, the underlying space—vacancies released during the annihilation of matter is what is referred to as dark matter, while volume energy is the candidate for dark energy. Radiation energy is the driving force for fluctuations in cosmic space. The ratio of volume energy to radiation energy is strictly 2:1. The former drives cosmic space expansion, while the latter causes disturbances in cosmic space fluctuations. The puzzling photon and electron double-slit interference experiments are the result of self-interference of quantum space fluctuations. In this sense, space and matter are identical. One can understand it as: matter is the curled space; space is the unfolded matter, both are composed of the underlying space.

5.2 Energy-Momentum Tensor of the Gravitational Field

We emphasize that the gravitational field is an energy field, the manifestation of the object's volume energy. The existence of solid matter deforms and bends the surrounding quantum space, and the elastic recovery of the curved space is the fundamental cause of gravity. In general relativity, Einstein established the gravitational field equations as

$$R_{\mu\nu} - \frac{1}{2} g_{\mu\nu} R = k_0 T_{\mu\nu} \tag{5-5}$$

and its deformation

$$R_{\mu}^{\nu} - \frac{1}{2} g_{\mu}^{\nu} R = k_0 T_{\mu}^{\nu} \tag{5-6}$$

In these equations, $R_{\mu\nu}$ and R_{μ}^{ν} represent the Ricci curvature tensor, $g_{\mu\nu}$, g_{μ}^{ν} is the metric tensor in four-dimensional space, R is the curvature scalar obtained by contracting the Ricci tensor, $T_{\mu\nu}$

and T_{μ}^{ν} represent the energy-momentum tensor density of matter, and k_0 is the proportional constant. The equation expresses that the curvature of space depends on the energy-momentum tensor of matter and its distribution. Since we emphasize that the gravitational field is the manifestation of volume energy, the volume energy density must be the energy component of the energy-momentum tensor. To clarify this fact, we will re-solve the metric solution for a spherically symmetric object. Let the interval in the four-dimensional free-space spherical polar coordinate system be

$$dS^2 = c^2dt^2 - dr^2 - r^2d\theta^2 - r^2\sin^2\theta d\varphi^2$$

$$= dx_0^2 - dr^2 - r^2d\theta^2 - r^2\sin^2\theta d\varphi^2 \qquad (5\text{-}7)$$

Under the influence of a spherically symmetric object, the curved space metric tensor still takes the Schwarzschild metric form, i.e.,

$$dS^2 = e^{N(r)}dx_0^2 - e^{L(r)}dr^2 - r^2d\theta^2 - r^2\sin^2\theta d\varphi^2 \qquad (5\text{-}8)$$

The metric tensor is

$$g_{\mu\nu} = \begin{bmatrix} e^{N(r)} & 0 & 0 & 0 \\ 0 & -e^{L(r)} & 0 & 0 \\ 0 & 0 & -r^2 & 0 \\ 0 & 0 & 0 & -r^2\sin^2\theta \end{bmatrix} \qquad (5\text{-}9)$$

Now, the unknown functions are $N(r)$ and $L(r)$. We will

solve for these unknown functions by solving equation (5-6). Einstein's field equation is a second-order partial differential equation, and solving for the metric tensor requires calculating the affine connection of the metric tensor. We omit the complex and lengthy calculation process and directly give the non-zero affine connection. Corresponding to the metric form in equation (5-9), the non-zero affine connection is

$$\Gamma^0_{01} = \Gamma^0_{10} = \frac{1}{2}N', \quad \Gamma^1_{00} = \frac{1}{2}N'e^{N-L}$$

$$\Gamma^1_{11} = \frac{1}{2}L', \quad \Gamma^1_{22} = -re^{-L}$$

$$\Gamma^1_{33} = -r\sin\theta e^{-L}, \quad \Gamma^2_{12} = \Gamma^2_{21} = 1/r$$

$$\Gamma^2_{33} = -\sin\theta\cos\theta, \quad \Gamma^3_{13} = \Gamma^3_{31} = 1/r$$

$$\Gamma^3_{23} = \Gamma^3_{32} = \cot\theta \qquad (5\text{-}10)$$

The Ricci tensor is derived from the Riemann curvature tensor, and the non-zero Riemann curvature tensor corresponding to equation (5-9) is

$$R^0_{101} = -\frac{1}{2}N'' + \frac{1}{4}N'L' - \frac{1}{4}(N')^2$$

$$R^0_{202} = -\frac{1}{2}rN'e^{-L}$$

$$R^1_{212} = \frac{1}{2}rL'e^{-L}$$

$$R^0_{303} = -\frac{1}{2}rN'e^{-L}\sin^2\theta$$

$$R_{313}^{1} = \frac{1}{2} r L' e^{-L} \sin^2\theta$$

$$R_{323}^{2} = \left(1 - e^{-L}\right)\sin^2\theta \tag{5-11}$$

The non-zero components of the contracted Ricci tensor are

$$R_0^0 = e^{-L}\left(-\frac{1}{2}N'' + \frac{1}{4}N'L' - \frac{1}{4}(N')^2 - \frac{N'}{r}\right)$$

$$R_1^1 = -e^{-L}\left(\frac{1}{2}N'' - \frac{1}{4}N'L' + \frac{1}{4}(N')^2 - \frac{L'}{r}\right)$$

$$R_2^2 = -\frac{1}{r^2}e^{-L}\left[1 + \frac{1}{2}r(N' - L')\right] + \frac{1}{r^2}$$

$$R_3^3 = -\frac{1}{r^2}e^{-L}\left[1 + \frac{1}{2}r(N' - L')\right] + \frac{1}{r^2} \tag{5-12}$$

Thus, the corresponding Ricci curvature scalar is

$$R = e^{-L}\left(-N'' + \frac{1}{2}N'L' - \frac{1}{2}(N')^2 + \frac{2}{r}(N' - L') - \frac{2}{r^2}\right) + \frac{2}{r^2} \tag{5-13}$$

For a spherically symmetric gravitational field, all components of the Ricci tensor, except for the diagonal ones, are zero, so the field equation simplifies to

$$R_\mu^\nu - \frac{1}{2}\delta_\mu^\nu R = k_0 T_\mu^\nu \tag{5-14}$$

Substituting equations (5-12) and (5-13) into the field equation (5-14) and simplifying, we obtain

$$R_0^0 - \frac{1}{2}R = -e^{-L}\left(\frac{L'}{r} - \frac{1}{r^2}\right) - \frac{1}{r^2}$$

$$R_1^1 - \frac{1}{2}R = e^{-L}\left(\frac{N'}{r} + \frac{1}{r^2}\right) - \frac{1}{r^2}$$

$$R_2^2 - \frac{1}{2}R = e^{-L}\left(\frac{N''}{2} - \frac{N'L'}{4} + \frac{(N')^2}{4} + \frac{(N'-L')}{2r}\right)$$

$$R_3^3 - \frac{1}{2}R = e^{-L}\left(\frac{N''}{2} - \frac{N'L'}{4} + \frac{(N')^2}{4} + \frac{(N'-L')}{2r}\right) \qquad (5\text{-}15)$$

With the above theoretical preparation, we proceed to solve for the unknown functions $N(r)$ and $L(r)$. Here, we emphasize that the assumption of zero energy-momentum tensor outside of matter in solving the field equations has a logical flaw. The core idea of general relativity is: the energy-momentum tensor of matter determines the curvature of four-dimensional space. A space without an energy-momentum tensor distribution is a flat, free space, and the metric tensor directly corresponds to equation (5-7) without the need to solve equation (5-9). Therefore, it is necessary to introduce the energy density of the gravitational field into the field equations. We point out that the gravitational field is the manifestation of volume energy. For a spherical object with mass m and nuclear density volume V_n, the energy density of the gravitational field is

$$E = 2G\left(\varepsilon_r^2 + \varepsilon_\theta^2 + \varepsilon_\varphi^2\right)$$

$$= 2G\left((\frac{-2m}{4\pi\rho_n r^3})^2 + (\frac{m}{4\pi\rho_n r^3})^2 + (\frac{m}{4\pi\rho_n r^3})^2 \right)$$

$$=2G\left(\frac{6m}{4\pi\rho_n r^3}\frac{m}{4\pi\rho_n r^3}\right)=6k\left(\frac{m}{4\pi\rho_n r^3}\frac{m}{4\pi\rho_n r^3}\right) \quad (5\text{-}16)$$

Einstein's field equation is universal and independent of the nuclear density volume of the gravitational source V_n. Therefore, we also need to know the energy density per unit nuclear density volume, denoted as E_ε. Noticing that $V_n = m/\rho_n$, we get

$$E_\varepsilon = \frac{E}{V_n} = 6k\left(\frac{m}{4\pi\rho_n r^3}\frac{1}{4\pi r^3}\right) \quad (5\text{-}17)$$

For the static field, all terms of the energy-momentum tensor are zero except for $T_0^{\ 0}$. Substituting equation (5-17) into the field equation, we get

$$R_0^0-\frac{1}{2}R=e^{-L}\left(-\frac{L'}{r}+\frac{1}{r^2}\right)-\frac{1}{r^2}=6k_0k\left(\frac{m}{4\pi\rho_n r^3}\frac{1}{4\pi r^3}\right)$$

$$R_1^1-\frac{1}{2}R=e^{-L}\left(\frac{N'}{r}+\frac{1}{r^2}\right)-\frac{1}{r^2}=0$$

$$R_2^2-\frac{1}{2}R=R_3^3-\frac{1}{2}R=e^{-L}\left(\frac{N''}{2}-\frac{N'L'}{4}+\frac{(N')^2}{4}+\frac{(N'-L')}{2r}\right)=0 \quad (5\text{-}18)$$

We subtract the second equation from the first equation above, obtaining

$$e^{-L}\left(\frac{L'+N'}{r}\right)=-6k_0k\left(\frac{m}{4\pi\rho_n r^3}\frac{1}{4\pi r^3}\right)=-k_0E_\varepsilon \quad (5\text{-}19)$$

As r becomes sufficiently large, the energy density $E_\varepsilon \to 0$, from equation (5-18), we can deduce that

$$N'(r) = -L'(r) \tag{5-20}$$

Thus,

$$N(r) = -L(r) + a$$

When r becomes sufficiently large, $e^N \to 1$, $e^{-L} \to 1$, so the constant $a = 0$. Hence,

$$N(r) = -L(r) \tag{5-21}$$

Now, replacing $N(r), N'(r)$ in equation (5-19) with $-L(r)$, $-L'(r)$, we have

$$e^N\left(\frac{N'+L'}{r}\right) = e^N\left(\frac{N'-N'_*}{r}\right) = e^N\frac{\delta(N')}{r}$$

$$= -k_0 E_\varepsilon \tag{5-22}$$

This represents the function and $(N'+L')$ in variational form. N'_* represents a function that is sufficiently close to N'. Now, let's solve equation (5-22).

We integrate both sides of equation (5-22), with the integration range from r to a point sufficiently far from the gravitational source r_a. Noticing that $\delta N(r) = N'(r)\delta r$, we get

$$\int_r^{r_a}\frac{e^{N(r)}}{r}\delta(N'(r)) = \int_r^{r_a}\frac{e^{N(r)}}{r}(\delta N(r))' = \int_r^{r_a}\frac{e^{N(r)}}{r}(N'(r))'\delta r$$

$$=\frac{e^{N(r)}}{r}N'(r)\Big|_r^{r_a}-\int_r^{r_a}\frac{\partial}{\partial r}\left(\frac{e^{N(r)}}{r}\right)N'(r)\delta r$$

$$=\frac{e^{N(r)}}{r}N'(r)\Big|_r^{r_a}-\int_r^{r_a}\frac{\partial}{\partial r}\left(\frac{e^{N(r)}}{r}\right)\delta N(r)$$

$$=\frac{e^{N(r)}}{r}N'(r)\Big|_r^{r_a}-\frac{\delta N(c)}{dr}\int_r^{r_a}\frac{\partial}{\partial r}\left(\frac{e^{N(r)}}{r}\right)dr \qquad (5\text{-}23)$$

In the last step, the mean value theorem of integration is applied. Since the variation of $\delta N(r)$ is infinitesimal, for any point c in the interval $[r, r_a]$, we can always make $\delta N(c) \to 0$, so the second term contributes zero to the integral. Therefore,

$$\int_r^{r_a}\frac{e^{N(r)}}{r}\delta(N'(r))=\frac{e^{N(r)}}{r}N'(r)\Big|_r^{r_a}=-\frac{e^{N(r)}}{r}N'(r) \qquad (5\text{-}24)$$

The right-hand side, E_ε, is the energy density of the gravitational field, which is excited by the unit nuclear density volume. Thus, the variation is δV_n, and the integral form of the right-hand side is

$$-k_0\int_{\delta V_n}E_\varepsilon\delta V_n=-6k_0k\int_{\delta V_n}\left(\frac{m}{4\pi\rho_n r^3}\frac{1}{4\pi r^3}\right)\delta V_n$$

$$=-6k_0k\int_r^{r_a}\left(\frac{m}{4\pi\rho_n r^3}\frac{1}{4\pi r^3}\right)(4\pi r^2)\delta r$$

$$=-\frac{6k_0km}{4\pi\rho_n}\int_r^{r_a}\frac{1}{r^4}\delta r=\frac{2k_0km}{4\pi\rho_n}\frac{1}{r^3}\bigg|_r^{r_a}$$

$$=-\frac{2k_0km}{4\pi\rho_n}\frac{1}{r^3}=-\frac{2k_0Gm}{r^3} \tag{5-25}$$

In this equation, k_0 represents the unknown coefficient of the gravitational field equation, k is the isotropic Young's modulus of quantum space, and G is the gravitational constant. Now, linking equations (5-24) and (5-25), we multiply both sides by,Then there is

$$-\frac{e^{N(r)}}{r}N'(r)=-\frac{2k_0Gm}{r^3} \tag{5-26}$$

r and integrate once more. The left-hand side becomes

$$-\int_r^{r_a}\left(e^{N(r)}N'(r)\right)dr=-e^{N(r)}\Big|_r^{r_a}=e^{N(r)}-1 \tag{5-27}$$

The right-hand side is

$$-2k_0Gm\int_r^{r_a}\frac{1}{r^2}dr=2kGm\frac{1}{r}\Big|_r^{r_a}=-\frac{2k_0Gm}{r} \tag{5-28}$$

Equating both sides, we obtain

$$e^{N(r)}=1-\frac{2k_0Gm}{r} \tag{5-29}$$

From $L(r)=-N(r)$, we know

$$e^{L(r)}=\frac{1}{1-2k_0Gm/r} \tag{5-30}$$

Substituting this conclusion into the third equation of (5-17), the equation holds. Equations (5-29) and (5-30) are the famous

Schwarzschild solution for a spherically symmetric gravitational field. Regarding the coefficient k_0, we will seek it from the variation in the frequency of gravitational field photons.

Let us assume a photon with frequency ν is emitted from gravitational potential U_r $(U_r = -Gm/r)$ towards gravitational potential $U_a(U_a \to 0)$. According to the mass-energy relation, the photon's dynamic mass is

$$\mu = \frac{E}{c^2} = \frac{h\nu}{c^2} \tag{5-31}$$

The work done by the photon as it moves from r to r_a is

$$W = 0 - \mu U_r = \frac{Gm\mu}{r} = \frac{Gmh\nu}{c^2 r} \tag{5-32}$$

Thus, the energy of the photon when it reaches r_a is

$$E_a = E - W = h\nu - \frac{Gmh\nu}{c^2 r}$$

$$= h\nu\left(1 - \frac{Gm}{c^2 r}\right) \tag{5-33}$$

Expressing the photon energy E_a in terms of frequency, we get

$$h\nu_a = h\nu\left(1 - \frac{Gm}{c^2 r}\right) \tag{5-34}$$

Equation (5-34) shows that, for the same wave packet, the

frequency decreases as it moves farther from the gravitational source. However, frequency is a measure of the clock rate in a gravitational field, and the frequency near the gravitational source is higher than the frequency in the outer field, indicating that time is slower near the gravitational source. Therefore, we rewrite equation (5-33) as

$$h\frac{1}{T_a} = h\frac{1}{T_r}\left(1-\frac{Gm}{c^2 r}\right) \tag{5-35}$$

Simplifying, we get

$$T_r = T_a\left(1-\frac{Gm}{c^2 r}\right) \tag{5-36}$$

From formula (5-36), when the time elapsed at r_a in the external field is dt, the time elapsed at r is dt_r .which is

$$dt_r = dt\left(1-\frac{Gm}{c^2 r}\right) \tag{5-37}$$

Now returning to the Schwarzschild metric form, and replacing the coordinate x^0 with ct, we get

$$\begin{aligned} c^2 dt_r^2 &= c^2 dt^2\left(1-\frac{2k_0 Gm}{r}\right) \\ &= c^2 dt^2\left(1-\frac{k_0 Gm}{r}\right)^2 \end{aligned} \tag{5-38}$$

Comparing this with equation (5-37), we conclude

$$k_0 = \frac{1}{c^2} \tag{5-39}$$

Substituting k_0 into the equation, the static solution for the spherically symmetric gravitational source Einstein field equation is

$$dS^2 = \left(1 - \frac{2Gm}{c^2 r}\right)c^2 dt^2 - \left(1 - \frac{2Gm}{c^2 r}\right)^{-1} dr^2 - r^2 d\theta^2 - r^2 \sin^2\theta d\varphi^2 \tag{5-40}$$

Equation (5-40) is the complete Schwarzschild solution. Here, we did not assume the energy density of the field outside matter is zero, nor did we use the weak field approximation to obtain an approximate solution. It is a direct result of the energy density function's variational calculation. This proves the universality of volume energy and the important conclusion that the elastic potential energy density of quantum space is a component of the gravitational field's energy-momentum tensor. Interestingly, substituting $k_0 = 1/c^2$ into equation (5-5), we get

$$R_{\mu\nu} - \frac{1}{2} g_{\mu\nu} R = \frac{1}{c^2} T_{\mu\nu} \tag{5-41}$$

Equation (5-41) indicates that Einstein's field equation can be understood as the wave equation of the gravitational field.

5.3 Dynamics of Cosmic Expansion

Regarding cosmic expansion, the significance of equation (5-4) is that it provides us with the dynamics of cosmic expansion. In the 1920s, American astronomer E.P. Hubble measured the redshift of spectra from known galaxies and derived the formula for the velocity of galaxies relative to the Milky Way as

$$V = Hd \tag{5-42}$$

In this equation, d represents the distance from a known galaxy to the center of the Milky Way, and H is Hubble's constant. Hubble's law, confirmed by astronomical observations, shows the universe is expanding. We believe that cosmic expansion originates from the thermonuclear reactions of stars. With a few assumptions, Einstein's mass-energy relation can provide a dynamical explanation for cosmic expansion.

We point out that the annihilation of matter releases vacancies, with nuclear density $\rho_n \to \rho_0$, rapidly expanding the geometric volume. The vacancies emitted by mass annihilation move at high speed through quantum space, and when they stop at a specific location in quantum space, they form part of space. In particular, the vacancies emitted by supernova explosions move at superluminal speeds. The fast particles observed during supernova explosions are precisely the vacancies emitted by matter annihilation. 67% of the total energy released by mass annihilation is converted into volume energy, and this enormous volume energy drives the expansion of the universe. As shown in Figure (5-1).

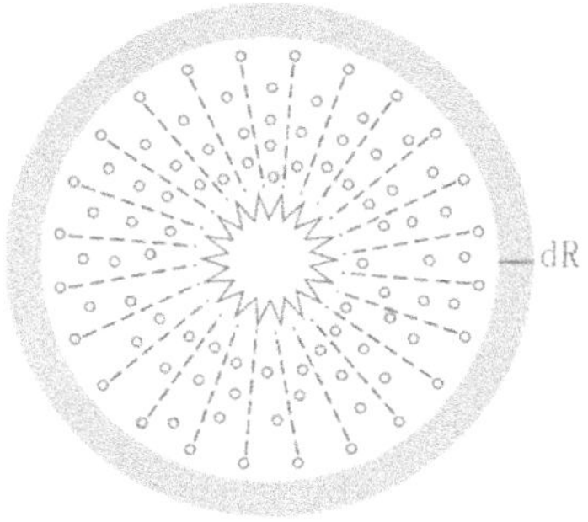

Figure (5-1) Schematic Diagram of Cosmic Expansion Dynamics.

The speed of cosmic expansion is determined by two factors: the mass deficit from the thermonuclear reactions of cosmic matter and the radius of the universe. Let at time t, the total mass of the universe's stars be M, and the mass deficit from stellar thermonuclear reactions in the time interval dt be dM. Generally, dM is proportional to M and time dt. Let H_0 be the proportional constant, then

$$dM = -H_0 M dt \tag{5-43}$$

In this equation, H_0 is referred to as the decay constant and is a sufficiently small quantity. Taking the Sun as an example, the energy radiated per second is $3.8\times10^{26}(J)$. According to the mass-energy relation, the mass deficit per second is $4.2\times10^{9}(kg)$. Substituting into equation (5-43), we know

$$H_0 = \frac{dM}{Mdt} = \frac{4.2\times10^{9}}{2.0\times10^{30}\times1} = 2.1\times10^{-21}(t^{-1})$$

By separating the variables and integrating equation (5-43), we get

$$M_t = Me^{-H_0 t} \tag{5-44}$$

In this equation, M_t is a function of the total mass of the stars as it changes with time, and M is the initial mass value corresponding to the observation time t. Suppose the universe is spherical, and the stellar system is uniformly distributed, with the radius of the universe at time t being R. Then the geometric expansion of the universe satisfies the following relationship:

$$\frac{dM}{\rho_0} = 4\pi R^2 dR \tag{5-45}$$

Substituting equation (5-43) into equation (5-45), simplifying, we get

$$V = \frac{dR}{dt} = \frac{H_0 M}{4\pi\rho_0 R^2} \tag{5-46}$$

This is another inverse square law. Equation (5-46) indicates that the speed of cosmic expansion is proportional to the total mass of the universe's stars, M, and inversely proportional to the instantaneous radius of the universe, R^2. In equation (5-46), V represents the geometric speed of cosmic expansion. Now, let us estimate the geometric speed of cosmic expansion. Based on the latest observational data, the radius of the observable universe is taken as 46.5 billion light-years, and the total mass of the observable universe's stars is $M = 2\times10^{53}(kg)$. Using the decay rate of solar material as a reference, we get

$$V = \frac{H_0 M}{4\pi\rho_0 R^2} = \frac{2.1\times10^{-21}\times2\times10^{53}}{4\times3.14\times2.08\times10^{-9}\times(4.65\times10^{10}\times9.46\times10^{15})^2}$$

$$= \frac{4.2\times10^{32}}{5.04\times10^{45}} = 8.34\times10^{-14}(m/t)$$

In this equation, ρ_0 is the mass density of quantum space. The result of the calculation shows that the cosmic stellar system consumes the mass equivalent of 420 Suns per second, with the geometric expansion speed being approximately $0.0834\,pm$ per second. In the calculation, we used the decay constant of the Sun, and now we know that the decay rate of massive stars is even higher, along with factors such as supernova explosions, gamma-ray bursts, black hole evaporation, etc. The average rate of mass annihilation in the universe may be far higher than the solar decay rate. Even if the decay rate increases by 10 orders of magnitude, the rate of cosmic geometric expansion will still be at the millimeter level. In this sense, the universe is stable for now.

However, Hubble's law states that the speed of cosmic expansion is proportional to its distance from us, and some conclusions even suggest that the universe is accelerating its expansion, even expanding faster than the speed of light. How can we explain this observational fact? Currently, the mainstream view is that the universe is an isolated system, and the total mass of stars cannot be created over time. Additionally, the mass deficit from stellar thermonuclear reactions decreases as the total mass of stars decreases. Therefore, under the condition that the amount of cosmic mass annihilation, dM, remains relatively stable, the universe itself does not have a driving force to accelerate its expansion. Hence, the cause of Hubble's law lies elsewhere.

We know that gravitational field radiation propagates from strong potential points to weak potential points and undergoes redshift. Suppose a galaxy emits a photon with frequency U_r from its gravitational potential v towards a distant point (sufficiently far), and the frequency of the photon received by the observer at the far point r_a is v'. The relationship between them is:

$$v' = v\left(1 - \frac{Gm}{c^2 r}\right) \tag{5-47}$$

The redshift is

$$\frac{\Delta v}{v} = \frac{v - v'}{v} = \frac{Gm}{c^2 r} = \frac{GM}{c^2}\frac{1}{r} \tag{5-48}$$

Equation (5-48) shows that the redshift is proportional to $(1/r)$. It is proportional to the gravitational potential of the star at the time of emission. As the universe expands, the speed of expansion is proportional to the mass deficit of the galaxy, dM , and inversely proportional to the instantaneous radius, r^2 . In the early universe, during the Big Bang, large eruptions, and annihilation, the mass, dM , was much greater than in the stable period, and the radius, r , was small, with a fast expansion speed. The further we move back in time, the smaller the radius of the universe. However, the mass of the cosmic galaxies remains stable, and the flux does not change. The gravitational field strength and potential become stronger as time moves forward. Therefore, equation (5-48) implies that the redshift is proportional to the time of the photon's travel through the universe, meaning it is proportional to the distance from the galaxy to us. This is the cause of Hubble's law.

In summary, the driving force for cosmic expansion comes from the volume energy converted from mass annihilation. If volume energy is referred to as dark energy, we can also say that the universe is expanding under the influence of dark energy. Dark energy is isotropic, insufficient to change the gravitational interactions between galaxies and planets, thus causing the expansion of the universe to manifest as the expansion of cosmic space.

We have strayed quite far from the main subject of this book, but using the derivation of Planck's constant analytical expressions, we have analyzed and discussed the elastic mechanical characteristics of space, foundational issues in quantum mechanics, and some basic questions of general relativity through volume energy analysis. Although we have diverged, this is still meaningful for interested readers.

By the way, the theory of quantized elastic space is an inheritance and development of the theory of elastic aether. Therefore, quantum space theory must address the sensitive issue of light's propagation through a medium and the coordination between light's propagation through a medium and special relativity. To this end, in 2007, we published a book titled On the Medium Propagation of Light (Northwest University Press), which rigorously demonstrated that relativistic physics, Lorentz transformations, are precisely the inevitable result of light's propagation through a medium.

References

[1] Xu Xudu, Chen Xiaoyu, Li Luo, eds. Physics Textbook (Volume 1 & 2). Beijing: Higher Education Press, 1989.

[2] Zhao Kaihua, Luo Weiyin, eds. Thermodynamics. Beijing: Higher Education Press, 1998.

[3] Zhao Kaihua, Luo Weiyin, eds. Quantum Mechanics. Beijing: Higher Education Press, 2001.

[4] Zhao Kaihua, Luo Weiyin, eds. Optics. Beijing: Higher Education Press, 2004.

[5] Wang Zhicheng, ed. Thermodynamics and Statistical Physics. Beijing: Higher Education Press, 2019.

[6] Zhu Bin, ed. Elastic Mechanics. Hefei: University of Science and Technology of China Press, 2008.

[7] Yang Guitong, ed. Brief Course in Elastic Mechanics. Beijing: Tsinghua University Press, 2013.

[8] Qin Fei, Wu Bin, eds. Fundamentals of Elasticity and Plasticity Theory. Beijing: Science Press, 2011.

[9] Zhang Zongsui, ed. Electrodynamics and Special Relativity. Beijing: Peking University Press, 2004.

[10] [Eng] Newton, translated by Wang Dike. Mathematical Principles of Natural Philosophy. Beijing: Peking University Press, 2006.

[11] [US] P.C. Bergman, translated by Zhou Qi, Hao Ping. Introduction to Relativity. Beijing: People's Education Press, 1961.

[12] [US] H.C. Vanian, [It] R. Ruffini, translated by Xiang Shouping, Feng Longlong. Gravitation and Spacetime. Beijing: Science Press, 2006.

[13] [US] Steve Adams, translated by Zhou Fuxin, Xuan Zhi, Shan Zhi. Physics of the 20th Century. Shanghai: Shanghai Scientific & Technical Publishers, 2006.

[14] [Rus] Kinzburg, translated by Wang Zhensong, Wang Ling, Luo Kun. Beijing: Science Press, 1987.

[15] [US] Particle Physics Special Group, translated by Shen Qixing, Wang Ping, Mao Huishun. Fundamental Particle Physics. Beijing: Science Press, 1992.

Beijing: Science Press, 1990.

[17] Guo Yiling, Shen Huijun, History of Physics. Beijing: Tsinghua University Press, 1993.

[18] Chen Dayou, On the Generation and Significance of Mass. Xi'an: Northwest University Press, 2007.

www.ingramcontent.com/pod-product-compliance
Lightning Source LLC
Chambersburg PA
CBHW021557210326
41599CB00010B/483
* 9 7 9 8 3 3 7 0 8 9 5 8 4 *